Advanced Information and Knowledge Processing

For other titles published in this series, go to
http://www.springer.com/series/4738

Animesh Adhikari · Pralhad Ramachandrarao ·
Witold Pedrycz

Developing Multi-database Mining Applications

Animesh Adhikari
Department of Computer Science
Smt. Parvatibal Chowgule
College
Margoa-403602
India
animeshadhikari@yahoo.com

Pralhad Ramachandrarao
Department of Computer Science &
Technology
Goa University
Goa-403206
India
pralhaad@rediffmail.com

Witold Pedrycz
Department of Electrical & Computer
Engineering
University of Alberta
9107 116 Street
Edmonton AB T6G 2V4
Canada
pedrycz@ece.ualberta.ca

ISBN 978-1-4471-2563-1 ISBN 978-1-84996-044-1(eBook)
DOI 10.1007/978-1-84996-044-1
Springer London Dordrecht Heidelberg New York

British Library Cataloguing in Publication Data
A catalogue record for this book is available from the British Library

Printed on acid-free paper

Springer is part of Springer Science+Business Media (www.springer.com)

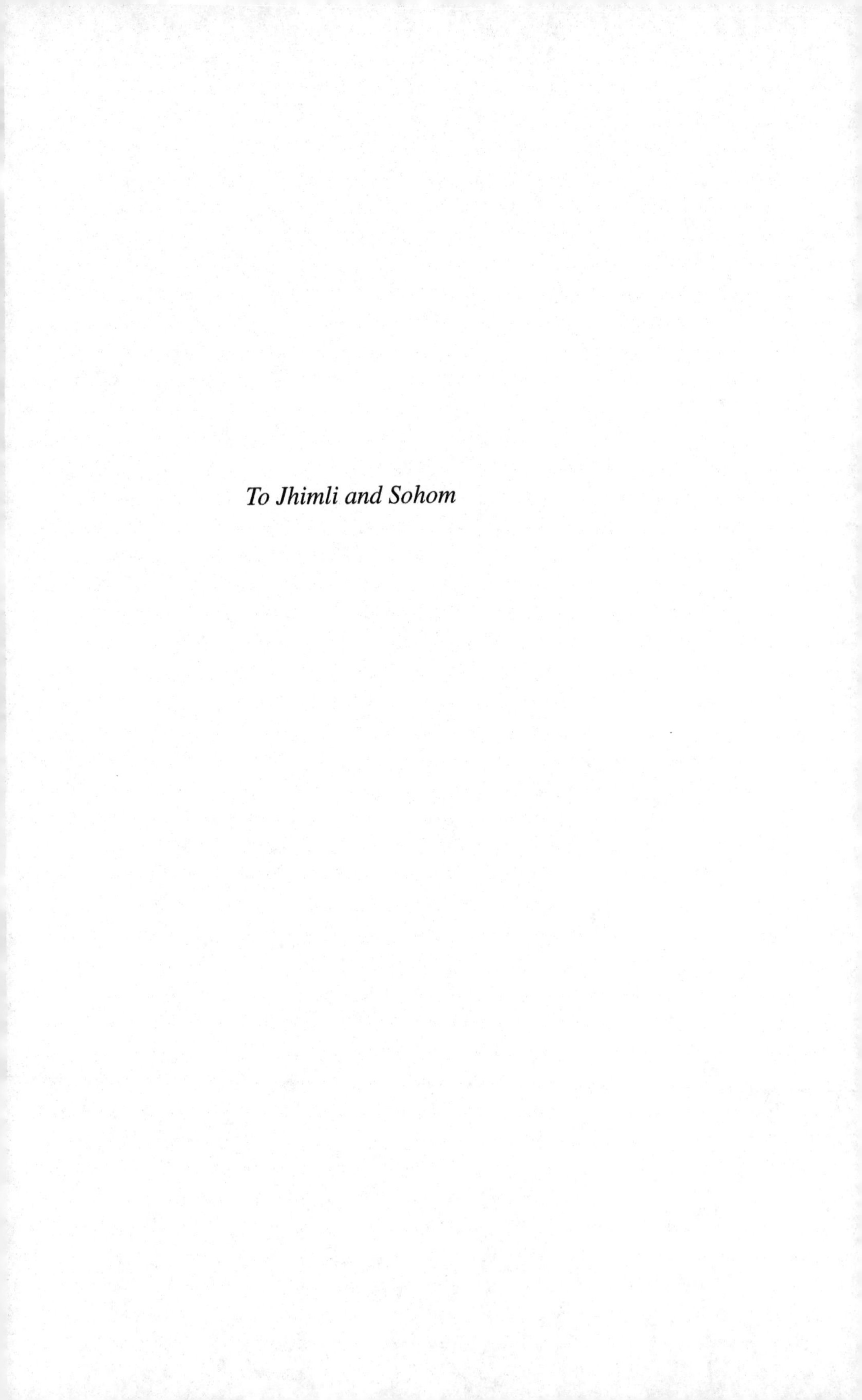

To Jhimli and Sohom

Contents

Chapter 1
Introduction

Many large organizations operate from multiple branches. Some of these branches collect data continuously. Thus, there are multi-branch organizations that possess multiple databases. Global decisions made by such an organization might be more appropriate if they are based on the data distributed over the branches. Moreover, the number of such applications is increasing over time. In this chapter, we discuss some of the major challenges encountered in multi-database mining that need to be dealt with. We discuss different issues of distributed data mining arising in this setting. In addition, we present three fundamental approaches to mining multiple large databases. We also elaborate on the recent developments that are taken place in this area. We provide a roadmap on how to develop an effective multi-database mining application and conclude the chapter by identifying some future research directions.

1.1 Motivation

With the advancement of science and technology, our civilization is changing at a faster rate. Also, rapid population growth has been another influential factor to support significant industrial growth and business activities. In addition, many countries across the globe are adopting slowly a liberal economic policy. Due to the influence of a number of such factors, some countries are experiencing rapid economic growth. As a result, the number of companies including those being multi-branch is increasing over time. In the recent time, the policy of merger and acquisition has become quite common. Many large companies operate from different branches located at different geographically distributed regions. Some of these branches are fully operational and collect transactional data on a continuous basis. As an example, consider shopping malls owned by a company. These malls are open at least 12 h a day. All the transactions made in a mall are stored locally. Thus, the company possesses multiple databases. It might be required to manage all these databases effectively for addressing different aspects of decision making especially if such problems need to be addressed at a global level.

Many important decisions could be based on the data distributed over the individual branches. Some global decisions might require an analysis of the entire data

A. Adhikari et al., *Developing Multi-database Mining Applications*, Advanced Information and Knowledge Processing, DOI 10.1007/978-1-84996-044-1_1,

distributed over the branches. The validity of the decisions would also depend on how effectively one can handle and comprehend relevant data at different branches. There exist some other categories of applications also where one would need to mine multiple large databases.

The domain of multi-database mining is increasing over time. The types and complexities of problems we encounter here are likely to increase in the future. Consider storing in a database of operational details of a single-cell organism. At minimum one would need to encode the following (Page and Craven 2003):

- *Genome*: DNA sequence and gene locations
- *Proteome*: the organism′s full complement of proteins, not necessarily a direct mapping from its genes
- *Metabolic pathways*: linked biochemical reactions involving multiple proteins, small molecules and protein-protein interactions
- *Regulatory pathways*: the mechanism by which the expression of some genes into proteins, such as transcription factors, influences the expression of other genes – includes protein-DNA interactions

In fact, such a database exists for part of what is known of the widely studied model organism E. coli-EcoCyc (Karp et al. 1997). Recording the diversity of data requires a rich and diversified relational schema with multiple, interacting relational tables. In fact, even recording one type of data, such as metabolic pathways, requires multiple relational tables because of the linked nature of pathways. It is not surprising that in this context multi-database mining (MDM) starts playing an essential role in reaching an effective goal. However, in this book we present studies based on multiple transactional databases.

Discovering knowledge from a large database is an interesting yet highly challenging issue. One of the visible challenges comes due to large size of a database. In many applications, multiple databases are required to be mined. No doubt that in these cases the challenges are increased manifold. In what follows, let us identify and discuss some of the major challenges one has to deal with.

- *Size of databases*: Some of the local databases could be large. Thus, the collection of all the branch databases is very large. A traditional data mining technique (Agrawal and Srikant 1994; Han et al. 2000) might take unreasonably large amount of time to process the collection of all databases present at individual branches. Sometimes, it might not be feasible to carry out the centralized data mining using a single computer. Another solution to this problem would be to employ parallel machines. This, unfortunately, might call for high investment on hardware and software. We need to make a thorough cost-benefit analysis before proceeding with such a decision. In some cases, it might not be an acceptable solution to the management of the company. Moreover, it might be difficult to find regional patterns when a traditional data mining technique is applied on the entire database. Thus traditional data mining techniques might not be the most suitable and fully recommended alternative in this situation.

- *Variety of data formats*: It might be possible that all the data sources come in different formats. We need to process them before proceeding with any data mining activity. Relevant data are required to be retained. Also, the definitions of data are required to be the same at every data source. Moreover, real-world data may be noisy. Thus, the preparation of data warehouse could be a significant task when handling multiple large databases.
- *Synthesis of non-local patterns*: The process of synthesizing non-local patterns is a challenging issue. In many cases, a pattern which is not reported from a local database is assumed as absent in that database. As a result, a synthesized non-local pattern then becomes approximate. There might exist a cascading, rippling effect on the decisions made on the basis of such approximate non-local patterns.
- *Limitations of exiting techniques of data mining*: Existing techniques for dealing with multiple large databases might not be satisfactory in all these situations. In Section 1.3, we discuss three important approaches to mining multiple large databases. We will also demonstrate that the existing multi-database mining techniques are not effective in all the situations.

In the subsequent chapters, we will address many design issues either in the context of a specific problem, or in general, for the purpose of developing effective multi-database mining applications.

The chapter is organized as follows. In Section 1.2, we provide an overview of mining distributed databases. In Section 1.3, we discuss existing approaches to mining multiple large databases. We discuss different applications of multi-database mining in Section 1.4. In Section 1.5, we present various issues on the development of effective multi-database mining applications. Finally, in Section 1.6 we identify some future research directions.

1.2 Distributed Data Mining

Distributed data mining (DDM) algorithms deals with mining multiple databases distributed over different geographical regions. In the last few years, researchers have started addressing problems where the databases stored at different places cannot be moved to a central storage area for variety of reasons. In multi-database mining, there are no such restrictions. Thus, distributed data mining could be considered as a special type of multi-database mining. Distributed data mining environment often comes with different distributed sources of computation. The advent of ubiquitous computing (Greenfield 2006), sensor networks (Zhao and Guibas 2004), grid computing (Wilkinson 2009), and privacy-sensitive multiparty data (Kargupta et al. 2003) present examples where centralization of data is either not possible, or at least not always desirable.

There is no doubt that ubiquitous computing could be the next wave of computing. We experienced the first wave of computing due to the excessive use of mainframes in both academia and industries. Each mainframe is shared by lots of people. Now we are in the personal computing era, person and machine face at each other uncomfortably across the desktop. Moreover, a person sometimes is needed to spend hours together to finish the task. It makes a person tiresome. Next comes ubiquitous computing, or the age of *calm* technology, when technology recedes quietly into the background of our lives. As opposed to the desktop paradigm, in which a single user consciously engages a single device for a specialized purpose, someone using ubiquitous computing engages many computational devices and systems simultaneously, in the course of ordinary activities, and may not necessarily even be aware that they are doing so.

There are many domains where distributed processing of data becomes a natural and scalable solution. Distributed wireless applications define one of such important domains. Consider an ad hoc wireless sensor network where different sensor nodes are monitoring some time-critical events. Central collection of data from every sensor node may create heavy traffic over the limited bandwidth offered by wireless channels and this may also drain a lot of power from the individual devices. Apart from the issue of power consumption, DDM over wireless networks also requires an application to run efficiently as many applications are time bound. The system might require to monitor and mine the on-board data stream generated by different sensors. Thus, centralization of databases is not desirable at all.

Many privacy-sensitive data mining adopt a distributed framework. The participating nodes exchange minimal amount information without transmitting raw data. Stolfo et al. (1997) designed JAM system for mining multiparty distributed sensitive data such as financial fraud detection. Distributed data in health care, finance, counter-terrorism and homeland defense often use sensitive data held by different parties. This comes into direct conflict with an individual's need and right to privacy. Yi and Zhang (2007) have proposed a privacy-preserving distributed association rule mining protocol based on a semi-trusted mixer model. The protocol can protect the privacy of each distributed database against the coalition up to $n-2$ other data sites or even the mixer if the mixer does not collude with any data site. Zhan et al. (2006) have proposed a secure protocol for multiple parties to collaboratively conduct association rule mining without disclosing their private data to each other or any other parties. Zhong (2007) has proposed algorithms for both vertically and horizontally partitioned data, with cryptographically strong privacy. The author has presented two algorithms for vertically partitioned data; one of them reveals only the support count and the other reveals nothing. Inan et al. (2007) have proposed methods for constructing the dissimilarity matrix of objects from different sites in a privacy preserving manner which can be used for privacy preserving clustering as well as database joins, record linkage and other operations that require pair-wise comparison of individual private data objects horizontally distributed to multiple sites.

Industry, science, and commerce fields often need to analyze very large databases maintained over geographically distributed sites by using the computational power of distributed systems. Grid can play a significant role in providing an effective computational infrastructure support for this kind of data mining. Similarly, the advent of multi-agent systems has brought us a new paradigm for the development of complex distributed applications. During the past decades, there have been several models and systems proposed to apply agent technology building distributed data mining. Through a combination of these two techniques, Luo et al. (2007) have investigated the different issues to build DDM on grid infrastructure and designed an agent grid intelligent platform as a testbed. Data mining algorithms and knowledge discovery processes are both compute and data intensive; therefore a grid can offer a computing and data management infrastructure for supporting decentralized and parallel data analysis. Congiusta et al. (2007) discussed how grid computing can be used to support distributed data mining.

In this book, we deal with multiple transactional databases that are not necessarily sensitive. In the following section, we discuss how the existing approaches dealt with multiple large databases.

1.3 Existing Multi-database Mining Approaches

In the following sections, we discuss three approaches to mining multiple large databases. In a distributed data mining environment, we may encounter different types of data. For example, stream data, geographical data, image data, transactional data are quite common. In this book, we deal with multiple transactional databases.

1.3.1 Local Pattern Analysis

Based on the number of data sources, patterns in multiple databases could be classified into three categories. They are local patterns, global patterns and patterns that are neither local nor global. A pattern based on a single database is called a local pattern. Local patterns are useful for local data analysis and decision making problems (Adhikari and Rao 2008b; Wu et al. 2005). On the other hand, global patterns are based on all the databases under consideration. They are useful for global data analyses (Adhikari and Rao 2008a; Wu and Zhang 2003). A convenient way to mine global patterns is to mine each local database, and then analyze all the local patterns to synthesize global patterns. This technique is simply called *local pattern analysis*. Zhang et al. (2003) designed local pattern analysis for the purpose of addressing various problems related to multiple large databases. Let us consider n branches of a multi-branch company. Also, let D_i be the database corresponding to i-th branch, $i = 1, 2, \ldots, n$. The essence of mining multiple databases using local pattern analysis could be explained using Fig. 1.1.

Let LPB_i be the local pattern base corresponding to D_i, $i = 1, 2, \ldots, n$. In multi-database environment, local patterns could be used in three ways by (i) Analyzing local data, (ii) Synthesizing non-local patterns, and (iii) Measuring relevant statistics for a decision making problems. Multi-database mining using local pattern analysis could be considered as an approximate method of mining multiple large databases. Thus, it might be required to enhance the quality of knowledge synthesized from multiple databases.

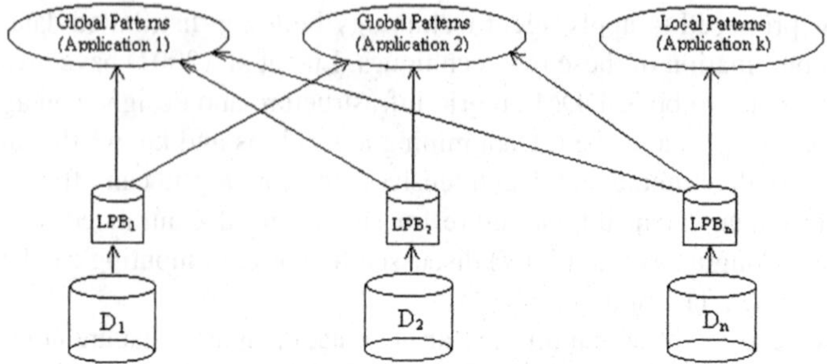

Fig. 1.1 Mining patterns in multiple databases using local pattern analysis

1.3.2 Sampling

In multi-database environment, the collection of all branch databases might be very large. Effective data analysis using a traditional data mining technique based on multi-gigabyte repositories has proven difficult. An approximate knowledge derived from large databases would be adequate for many decision support applications. Such applications could be advantageous to offer quick support in decision-making processes. In these cases, one could tame multiple large databases by sampling (Babcock et al. 2003). For instance, a commonly used technique for approximate query answering is sampling (Cochran 1977). If an itemset is frequent in a large database then it is likely that the itemset is also frequent in a sample data. Thus, one could analyze approximately the database by analyzing the frequent itemsets in a representative sample data. A combination of sampling and local pattern analysis could be a useful technique for mining multiple databases for addressing many decision support applications.

1.3.3 Re-mining

For the purpose of mining multiple databases, one could apply partition algorithm proposed by Savasere et al. (1995). The algorithm is designed for mining a very large database by partitioning. The algorithm works as follows. It scans the database twice. The database is divided into disjoint partitions, where each partition is small enough to fit in memory. In a first scan, the algorithm reads each partition and computes locally frequent itemsets in each partition using apriori algorithm (Agrawal and Srikant 1994). In the second scan, the algorithm counts the supports of all locally frequent itemsets toward the complete database. In this case, each local database could be considered as a partition. Though partition algorithm mines frequent itemsets in a database exactly, it might be an expensive solution to mining multiple large databases, since each database is required to be scanned twice. During the time of the second scanning, all the local patterns obtained at the first scan are analyzed. Thus, partition algorithm used for mining multiple databases could be considered as another type of local pattern analysis.

1.4 Applications of Multi-database Mining

Multi-database mining has been recently recognized as an important area of research in data mining. We discuss here a few applications of multi-database mining.

Kum et al. (2006) have proposed ApproxMAP algorithm, to mine approximate sequential patterns, called consensus patterns, from large sequence databases in two steps. First, sequences are organized into similarity groups, called clusters. Then, consensus patterns are mined directly from each cluster through multiple alignments.

Enterprise applications usually involve huge, complex, and persistent data to work on, together with business rules and processes. In order to represent, integrate, and use the information coming from huge, distributed, multiple sources, Hu and Zhong (2006) have presented a conceptual model with dynamic multi-level workflows corresponding to a mining-grid centric multi-layer grid architecture, for multi-aspect analysis in building an e-business portal on the Wisdom Web. The authors have showed that this integrated model would help to dynamically organize status-based business processes that govern enterprise application integration.

A multi-domain sequential pattern is a sequence of events whose occurrence time is within a pre-defined time window. Given a set of sequence databases across multiple domains, Peng and Liao (2009) have aimed at mining multi-domain sequential patterns.

A multi-branch company is often interested in high-frequency rules because they are supported by most of its branches for corporate profitability. Wu and Zhang (2003) have proposed a weighting model for synthesizing high-frequent association rules from different data sources.

To reduce the search cost in the data from all databases, we need to identify which databases are most likely relevant to a data mining application. For this purpose, Wu et al. (2005) have proposed an algorithm for selecting relevant databases.

Ratio rules are aimed at capturing the quantitative association knowledge. Yan et al. (2006) have extended this framework to mining ratio rules from distributed and dynamic data sources. Authors have proposed an integrated method to mining ratio rules from distributed and changing data sources, by first mining the ratio rules from each data source separately through a novel robust and adaptive one-pass algorithm, and then integrating the rules of each data source in a simple probabilistic model.

Zhang et al. (2009) have proposed a nonlinear method using kernel estimation for mining global patterns in multiple databases. A global exceptional pattern describes interesting individuality of few branches. Therefore, it is interesting to identify such patterns. Adhikari and Rao (2007), Zhang et al. (2004a) have introduced different strategies for identifying global exceptional patterns in multiple databases.

Principal component analysis (PCA) is frequently used for constructing the reduced representation of the data. The method often reduces the dimensionality of the original data by a large factor and constructs features that capture the maximally varying directions in the data. Kargupta et al. (2000) have proposed a technique of computing the collective principal component analysis from heterogeneous sites.

Biological databases contain a wide variety of data types, often with rich relational structure. Consequently multi-relational data mining techniques frequently are applied to biological data. Page and Craven (2003) have presented several applications of multi-relational data mining to biological data, taking care to cover a broad range of multi-relational data mining techniques. The field of bioinformatics is expanding rapidly. In this field large multiple as well as complex relational tables are dealt with frequently. Wang et al. (2005) present various techniques in biological data mining and data management. The book also includes preprocessing tasks such as data cleaning and data integration as applied to biological data.

A general discussion on multi-database mining, applications, various issues and challenges can be found in Zhang et al. (2004b). Kargupta et al. (2004) have edited a book containing various issues on distributed data mining.

1.5 Improving Multi-database Mining

One could mine multiple databases using traditional data mining techniques or consider the use of non-traditional techniques. Some examples of traditional data mining techniques are apriori algorithm (Agrawal and Srikant 1994), FP-growth algorithm (Han et al. 2000), and P-tree algorithm (Coenen et al. 2004). For applying a traditional data mining technique, one needs to amass all the databases together. Thus, the collection of branch databases could be then thought as a single source of data. In virtue of the process, the patterns extracted are exact. Thus, no improvement of patterns (output) is required. But, it might be possible to improve different traditional data mining algorithms with respect to time complexity, space complexity, and other parameters of different mining algorithms. Though these are interesting topics, in this book we will not be concerned about these issues. Some examples of non-traditional data mining techniques that could be used for mining multiple databases are partition algorithm (Savasere et al. 1995), local pattern analysis (Zhang et al. 2003) and sampling technique (Babcock et al. 2003). In Section 1.3, we have noted several drawbacks of each of the non-traditional data mining techniques. We propose various strategies for improving multi-database mining applications. Some improvements are general in nature, while others are more specific. The efficiency of a multi-database application could be enhanced by choosing a better multi-database mining model, a better pattern synthesizing technique, a better pattern representation technique and a better algorithm for solving the problem. This book illustrates each of these issues either in the context of a specific problem, or in some general setting. It does not discuss an efficient implementation of different algorithms, since the topic has been well studied.

1.5.1 Various Issues of Developing Effective Multi-database Mining Applications

It might be possible to improve a multi-database mining application, if we critically and constructively analyze each step of the development process. In what follows, we provide a brief description of the remaining chapters, and highlight how they become instrumental in building effective multi-database mining applications.

In Chapter 2, we present an extended model for synthesizing global patterns from local patterns present in different databases. We use this model to show how one could systematically develop different multi-database mining applications using local pattern analysis. For example, one could mine a specific type of global patterns in multiple databases. In this context, we have presented the notion of heavy association rule in multiple databases. Also, we have presented an algorithm for synthesizing heavy association rules in multiple databases. In addition, the notion of exceptional association rule in multiple databases is presented, and an extension is made to this algorithm to notify whether a heavy association rule is high-frequent or exceptional. We present experimental results in case of three real-world databases. Also, we provide a comparative analysis of the proposed algorithm with the existing algorithms.

Effective data analysis with multiple databases requires highly accurate patterns. But local pattern analysis might extract low quality of patterns from multiple databases. Thus, it becomes necessary to improve mining multiple databases. In Chapter 3, we present a new technique of mining multiple databases in which each local database is mined using a traditional data mining technique in a particular order for synthesizing global patterns. The technique improves significantly the quality of synthesized global patterns. We conduct experiments on both real and synthetic databases to quantify the effectiveness of the proposed technique.

Many important decisions are based on a set of specific items called the *select items*. Thus, the analysis of select items in multiple databases is an important task. In Chapter 4, we discuss how one could extract patterns related to select items exactly from multiple large databases. Thus, we present a model of mining global patterns of select items from multiple databases. Then, a measure of overall association between two items in a database is proposed. Finally, an algorithm is designed based on overall association between two items in a database for the purpose of grouping the frequent items in local databases. Each group contains a select item called the *nucleus item*, and the group grows being centered around the nucleus item. Experimental results are provided using both real-world and synthetic databases.

Multi-database mining using local pattern analysis could be considered as an approximate method of mining multiple large databases. Thus, it might be required to enhance the quality of knowledge synthesized from multiple databases. Also, many decision-making applications are directly based on the available local patterns in different databases. The quality of synthesized knowledge/decision based on local patterns in different databases could be enhanced by incorporating more local patterns in the knowledge synthesizing/processing activities. Thus, the available local patterns play a crucial role in building efficient multi-database mining applications. In Chapter 5, we represent patterns in condensed form by employing a coding called antecedent consequent pair (ACP) coding. It allows us to consider more local patterns by lowering further the user inputs, like minimum support and minimum confidence. The proposed coding enables more local patterns participate in the knowledge synthesizing / processing activities and thus, the quality of synthesized knowledge based on local patterns in different databases gets enhanced significantly at a given pattern synthesizing algorithm and computing resource.

In Chapter 6, we present two measures of similarity between a pair of databases. Also, we present an algorithm for clustering a set of databases. We have enhanced the efficiency of the clustering process using several strategies such as reducing the execution time of clustering algorithm, using more appropriate similarity measure, and efficiently storing frequent itemsets space.

1.6 Experimental Settings

We have carried out several experiments to study the effectiveness of the proposed approaches in different chapters. For Chapters 2, 4, 5 and 6, all the experiments have been realized on a 1.6 GHz Pentium processor with 256 MB of memory using Visual C++ (version 6.0) software. For Chapter 3, all the experiments have been implemented on a 2.8 GHz Pentium D dual core processor with 512 MB of memory using Visual C++ (version 6.0) compiler.

1.7 Future Directions

Multi-database mining is also applicable in other domains. In Section 1.1, we have cited an example where multi-relational data mining is applied quite often in the field of bioinformatics. In this book, we have confined our discussion on mining multiple large transactional databases. We will discuss different strategies to improve multi-database mining applications in the context of multiple large transactional databases. Similar strategies could also be adopted to handle multiple databases in other domains.

World Wide Web (WWW) is a large distributed repository of data. Su et al. (2006) have proposed a logical framework for identifying quality knowledge from different data sources. Various studies on WWW data might dominate future studies.

The popularity of the Internet as well as the availability of powerful computers and high-speed network technologies as low-cost commodity components is changing the way we use computers today. These technology opportunities have led to the possibility of using distributed computers as a single, unified computing resource, leading to what is popularly known as Grid computing (Foster and Kesselman 1999). Clusters and grids of workstations provide available resources for data mining processes. To exploit these resources, new distributed algorithms are necessary, particularly concerning the way to distribute data and to use this partition. Fiolet and Toursel (2007) have presented a clustering algorithm known as distributed progressive clustering, for providing an "intelligent" distribution of data on grids. Cluster and grid computing will be playing a dominant role in the next generation of computing.

In a distributed environment, a large database could be fragmented vertically and/or horizontally. This might bring additional complexities for mining patterns in multiple large databases. Agrawal and Shaffer (1999) introduced a parallel version of apriori algorithm.

Distributed data mining for wireless applications is another active area of multi-database mining. Challenges here are somewhat different from that of classical multi-database mining. Bandwidth limitation is one of the major constraints in this domain. There are other constraints, such as power consumption. The next generation algorithms will have to deal with these important constraints.

Data privacy is likely to remain an important issue in data mining research and application. The field of privacy-preserving data mining has started recently. Da Silva and Klusch (2006) have proposed KDEC-S algorithm for distributed data clustering, which is shown to provide mining results while preserving confidentiality of original data. Stankovski et al. (2008) have designed *DataMiningGrid* system to meet the requirements of modern and distributed data mining scenarios. Based on the Globus Toolkit and other open technology and standards, the *DataMiningGrid* system provides tools and services facilitating the grid-enabling of data mining applications without any intervention on the application side. In future, the concepts and various issues will get formalized. More privacy-preserving algorithms are likely to appear as more applications on privacy-sensitive data are likely to emerge in the future.

Multi-agent systems (MAS) offer architecture for distributed problem solving. DDM algorithms focus on one class of such distributed problem solving tasks, analysis and modeling of distributed data. Da Silva et al. (2005) offer a perspective on DDM algorithms in the context of multi-agents systems. It discusses broadly the connection between DDM and MAS. In future, many DDM algorithms are likely to come in association with MAS.

With the increasing popularity of object-oriented database systems in advanced database applications, it is also important to study the data mining methods in object-oriented data. Han et al. (1998) investigated issues on generalization-based data mining in object-oriented databases considering three crucial aspects: (1) generalization of complex objects, (2) class-based generalization and (3) extraction of different kinds of rules. The authors proposed an object cube model for class-based generalization, on-line analytical processing and data mining. Various issues of multiple object-oriented databases deserve to be investigated.

Clinical laboratory databases are among the largest generally accessible, detailed records of human phenotype in disease, they will likely have an important role in future studies designed to tease out associations between human gene expression and the presentation and progression of disease. Multi-database mining will be playing an important role in this area (Siadaty and Harrison 2008).

The dramatic increase in the availability of massive, complex data from various sources is creating computing, storage, communication, and human-computer interaction challenges for data mining. Providing a framework to better understand these fundamental issues, Kargupta et al. (2008) have surveyed promising approaches to data mining problems that span an array of disciplines. In the coming years, we will witness more applications of multi-databases mining. We need to prepare ourselves to tackle various issues and problems related to mining multiple large databases.

References

Adhikari A, Rao PR (2007) Synthesizing global exceptional patterns in multiple databases. In: Proceedings of the 3rd Indian International Conference on Artificial Intelligence, pp. 512–531

Adhikari A, Rao PR (2008a) Synthesizing heavy association rules from different real data sources. Pattern Recognition Letters 29(1): 59–71

Adhikari A, Rao PR (2008b) Efficient clustering of databases induced by local patterns. Decision Support Systems 44(4): 925–943

Agrawal R, Shafer J (1999) Parallel mining of association rules. IEEE Transactions on Knowledge and Data Engineering 8(6): 962–969

Agrawal R, Srikant R (1994) Fast algorithms for mining association rules. In: Proceedings of International Conference on Very Large Data Bases, Santiago, Chile, pp. 487–499

Babcock B, Chaudhury S, Das G (2003) Dynamic sample selection for approximate query processing. In: Proceedings of ACM SIGMOD Conference Management of Data, New York, pp. 539–550

Cochran WG (1977) Sampling techniques. Third edition, Wiley, New York

Coenen F, Leng P, Ahmed S (2004) Data structure for association rule mining: T-trees and P-trees. IEEE Transactions on Knowledge and Data Engineering 16(6):774–778

Congiusta A, Talia D, Trunfio P (2007) Service-oriented middleware for distributed data mining on the grid. Journal of Parallel and Distributed Computing 68(1): 3–15

Da Silva JC, Giannellab C, Bhargava R, Kargupta H, Klusch M (2005) Distributed data mining and agents. Engineering Applications of Artificial Intelligence 18(7): 791–807

Da Silva JC, Klusch M (2006) Inference in distributed data clustering. Engineering Applications of Artificial Intelligence 19(4): 363–369

Fiolet V, Toursel B (2007) A clustering method to distribute a database on a grid. Future Generation Computer Systems 23(8): 997–1002

Foster I, Kesselman C (eds.) (1999) The Grid: Blueprint for a future computing infrastructure. Morgan Kaufmann, San Francisco

Greenfield A (2006) Everyware: The Dawning Age of Ubiquitous Computing. First edition, New Riders Publishing, Indianapolis, IN

Han J, Nishio S, Kawano H, Wang W (1998) Generalization-based data mining in object-oriented databases using an object cube model. Data and Knowledge Engineering 25(1–2): 55–97

Han J, Pei J, Yiwen Y (2000) Mining frequent patterns without candidate generation. In: Proceedings of ACM SIGMOD Conference on Management of Data, pp. 1–12

Hu J, Zhong N (2006) Organizing multiple data sources for developing intelligent e-business portals. Data Mining and Knowledge Discovery 12(2–3): 127–150

Inan A, Kaya SV, Saygın Y, Savas E, Hintoglu AA, Levi A (2007) Privacy preserving clustering on horizontally partitioned data. Data and Knowledge Engineering 63(3): 646–666

Kargupta H, Han J, Yu PS, Motwani R, Kumar V (2008) Next Generation of Data Mining. CRC Press, Bocca Raton

Kargupta H, Huang W, Krishnamurthy S, Park B, Wang S (2000) Collective PCA from distributed and heterogeneous data. In: Proceedings of the Fourth European Conference on Principles and Practice of Knowledge Discovery in Databases, Springer Verlag, pp. 452–457.

Kargupta H, Joshi A, Sivakumar K, Yesha Y (2004) Data Mining: Next Generation Challenges and Future Directions. MIT/AAAI Press, Cambridge, MA

Kargupta H, Liu K, Ryan J (2003) Privacy sensitive distributed data mining from multi-party data. In: Proceedings of Intelligence and Security Informatics, Springer-Verlag, pp. 336–342.

Karp P, Riley M, Paley S, Pellegrini-Toole A (1997) EcoCyc: Electronic encyclopedia of E. coli genes and metabolism. Nucleic Acids Research, 25(1), 43–50

Kum H-C, Chang HC, Wang W (2006) Sequential pattern mining in multi-databases via multiple alignment. Data Mining and Knowledge Discovery 12(2–3): 151–180

Luo J, Wang M, Hu J, Shi J (2007) Distributed data mining on Agent Grid: Issues, platform and development toolkit. Future Generation Computer Systems 23(1, 1): 61–68

Page D, Craven M (2003) Biological applications of multi-relational data mining. SIGKDD Explorations 5(1): 69–79

Peng W-C, Liao Z-X (2009) Mining sequential patterns across multiple sequence databases. Data & Knowledge Engineering 68(10): 1014–1033

Savasere A, Omiecinski E, Navathe S (1995) An efficient algorithm for mining association rules in large databases. In: Proceedings of the 21st International Conference on Very Large Data Bases, pp. 432–443

Siadaty MS, Harrison Jr JH (2008) Multi-database mining. Clinics in Laboratory Medicine 28(1): 73–82

Stankovski V, Swain M, Kravtsov V, Niessen T, Wegener D, Kindermann J, Dubitzky W (2008) Grid-enabling data mining applications with DataMiningGrid: An architectural perspective. Future Generation Computer Systems 24(4): 259–279

Stolfo S, Prodromidis AL, Chan PK (1997) JAM: Java agents for meta-learning over distributed databases. In: Proceedings of Third International Conference on Knowledge Discovery and Data Mining, pp. 74–81

Su K, Huang H, Wu X, Zhang S (2006) A logical framework for identifying quality knowledge from different data sources. Decision Support Systems 42(3): 1673–1683

Wang JT, Zaki MJ, Toivonen HT, Shasha DE (2005) Data Mining in Bioinformatics. Springer, London/New York

Wilkinson (2009) Grid computing: Techniques and applications, CRC Press, Boca Raton

Wu X, Zhang S (2003) Synthesizing high-frequency rules from different data sources. IEEE Transactions on Knowledge and Data Engineering 14(2): 353–367

Wu X, Zhang C, Zhang S (2005) Database classification for multi-database mining. Information Systems 30(1): 71–88

Yan J, Liu N, Yang Q, Zhang B, Cheng Q, Chen Z (2006) Mining adaptive ratio rules from distributed data sources. Data Mining and Knowledge Discovery 12 (2–3): 249–273

Yi X, Zhang Y (2007) Privacy-preserving distributed association rule mining via semi-trusted mixer. Data and Knowledge Engineering 63(2): 550–567

Zhan J, Matwina S, Chang LW (2006) Privacy-preserving collaborative association rule mining. Journal of Network and Computer Applications 30(3): 1216–1227

Zhang C, Liu M, Nie W, Zhang S (2004a) Identifying global exceptional patterns in multi-database mining. IEEE Computational Intelligence Bulletin 3(1): 19–24

Zhang S, Wu X, Zhang C (2003) Multi-database mining. IEEE Computational Intelligence Bulletin 2(1): 5–13

Zhang S, You X, Jin Z, Wu X (2009) Mining globally interesting patterns from multiple databases using kernel estimation. Expert Systems with Applications 36(8): 10863–10869

Zhang S, Zhang C, Wu X (2004b) Knowledge discovery in multiple databases. Springer, New York

Zhao F, Guibas L (2004) Wireless Sensor Networks: An Information Processing Approach. Morgan Kaufmann, San Francisco

Zhong S (2007) Privacy-preserving algorithms for distributed mining of frequent itemsets. Information Sciences 177(2): 490–503

Chapter 2
An Extended Model of Local Pattern Analysis

The model of local pattern analysis provides sound solutions to many multi-database mining problems. In this chapter, we will discuss different types of extreme association rules in multiple databases viz., heavy association rule, high-frequency association rule, low-frequency association rule and exceptional association rule. Also, we show how one can apply the model of local pattern analysis more systematically and effectively. For this purpose, we introduce an extended model of local pattern analysis. We apply the extended model to mine heavy association rules in multiple databases. Also, we justify why the extended model works more effectively. We develop an algorithm for synthesizing heavy association rule in multiple databases. Furthermore, we show that the algorithm identifies whether a heavy association rule is high-frequency rule or exceptional rule. We have provided experimental results obtained for both synthetic and real-world datasets and carried out detailed error analysis. Furthermore, we bring a detailed comparative analysis by contrasting the proposed algorithm with some of those reported in the literature. This analysis is completed by taking into consideration the criteria of execution time and average error.

2.1 Introduction

In the previous chapter, we have discussed limitations of using a conventional data mining technique for mining multiple large databases. Also we have discussed challenges involved in mining multiple large databases. In many decision support applications, an approximate knowledge stemming from multiple large databases might result in significant savings when being used in decision-making. Hence the model of local pattern analysis (Zhang et al. 2003) used for mining multiple large databases can constitute a viable solution. In this chapter, we show how one can apply the model of local pattern analysis in a systematic and efficient manner for mining non-local patterns in multiple databases.

For mining multiple large databases, careful preparation of data collected at the respective branches is of significant importance. In fact, data preparation can be divided into several sub-tasks, so that it makes the overall data mining easy to perform. We divide the overall data mining task into a hierarchy of sub-tasks to be

A. Adhikari et al., *Developing Multi-database Mining Applications*, Advanced Information and Knowledge Processing, DOI 10.1007/978-1-84996-044-1_2,

performed at each branch, and finally an application could be developed using local patterns at different branch databases. A non-local application might aim at mining non-local interesting patterns in multiple databases, or making a non-local decision based on findings realized in multiple databases. For determining a solution to the latter problem, sometimes we need to compute appropriate statistics based on the patterns discovered in multiple databases. An appropriate statistic then enables us to take such non-local decisions. For applying the extended model of mining multiple large databases, we have synthesized a specific type of global patterns in multiple databases. In Section 2.2, we discuss some interesting types of patterns in multiple databases.

The rest of the chapter is organized as follows. We discuss some "extreme" types of pattern (Section 2.2). In Section 2.3, we present an extended model of local pattern analysis. We present an application of the extended model in Section 2.4. Finally, some conclusions are provided in Section 2.5.

2.2 Some Extreme Types of Association Rule in Multiple Databases

The analysis of relationships among variables is a fundamental task positioned at the heart of many data mining problems. Mining association rules has received a lot of attention in the data mining community. For instance, an association rule expresses how the purchase of a group of items, called an *itemset*, affects the purchase of another group of items. Association rule mining is based on two measures quantifying the quality of the rules, that is support (*supp*) and confidence (*conf*) see Agrawal et al. (1993). An association rule r in database DB can be expressed symbolically as $X \rightarrow Y$, where X and Y are two itemsets in database DB. It expresses an association between the itemsets X and Y, called the antecedent and consequent of r, respectively. The meaning attached to this type of implication could be clarified as follows. If the items in X are purchased by a customer then the items in Y are likely to be purchased by the same customer at the same time. The interestingness of an association rule could be expressed by its support and confidence. Let E be a Boolean expression defined on the items in DB. Support of E in DB is defined as the fraction of transactions in DB such that the Boolean expression E is true for each of these transactions. We denote the support of E in DB as $supp_{\mathrm{a}}(E, DB)$. Then the support and confidence of association rule r could be expressed as follows:

$$supp_{\mathrm{a}}(r, DB) = supp_{\mathrm{a}}(X \cap Y, DB), \text{ and}$$
$$conf_{\mathrm{a}}(r, DB) = supp_{\mathrm{a}}(X \cap Y, DB)/supp_{\mathrm{a}}(X, DB)$$

Later, we will be dealing with synthesized support and synthesized confidence of an association rule. Thus, it is required to differentiate between actual support/confidence with synthesized support/confidence of an association rule. The subscript a used in the notation of support/confidence refers to the actual support/confidence of an association rule. On the other hand, the subscript s in the notation of support/confidence refers to synthesized support/confidence of an association rule. A synthesized support/confidence of an association rule might depend on the technique applied for synthesizing support/confidence. We will introduce and discuss a technique for synthesizing support and confidence of an association rule in multiple databases. We say that an association rule r in database DB is *interesting* if

$$supp_{\mathrm{a}}(r, DB) \geq minimum\ support(\alpha), \text{ and}$$
$$conf_{\mathrm{a}}(r, DB) \geq minimum\ confidence(\beta)$$

The values of the parameters α and β are user-defined. The collection of association rules extracted from a database for the given values of α and β is called a *rulebase*.

An association rule in multiple databases becomes more interesting if it possesses higher support and higher confidence. This type of association rule is called heavy association rules (Adhikari and Rao 2008). Sometimes the number of times an association rule gets reported from local databases becomes an interesting issue. In the context of multiple databases, an association rule is called high-frequency rule (Wu and Zhang 2003) if it is extracted from many databases. In this context an association rule is called low-frequency rule (Adhikari and Rao 2008) if it is extracted from a few databases. Some association rules possess high support but have been extracted from a few databases only. These association rules are called exceptional association rules (Adhikari and Rao 2008). Many corporate decisions could be influenced by these types of extreme association rules in multiple databases. Thus, it is important to mine them. In what follows, we define formally heavy association rule, high-frequency association rule, low-frequency association rule, and exceptional association rule in multiple databases.

Consider a large company with transactions originating from n branches. Let D_i be the database corresponding to the i-th branch of this multi-branch company, $i = 1, 2, \ldots, n$. Furthermore let D be the union of all branch databases. First, we define a heavy association rule in a single database. Afterwards, we define a heavy association rule in multiple databases.

Definition 2.1 *An association rule* r *in database* DB *is heavy if* $\mathrm{supp}_a(\mathrm{r, DB}) \geq \mu$, *and* $\mathrm{conf}_a(\mathrm{r, DB}) \geq \nu$, *where* μ $(> \alpha)$ *and* ν $(> \beta)$ *are the user-defined thresholds of high-support and high-confidence for identifying heavy association rules in* DB, *respectively.*

If an association rule is heavy in a local database then it might not be heavy in D. An association rule in D might have different statuses in different local databases. For example, it might be a heavy association rule, or an association rule, or a suggested association rule (defined later), or absent in a local database. Thus, we need to synthesize an association rule for determining its overall status in D. The method of synthesizing an association rule is discussed in Section 2.4.2. After synthesizing an association rule, we get its synthesized support and synthesized confidence in D. Let $supp_{\mathrm{s}}(r, DB)$ and $conf_{\mathrm{s}}(r, DB)$ denote synthesized support and synthesized confidence of association rule r in DB, respectively. A heavy association rule in multiple databases is defined as follows:

Definition 2.2 *Let* D *be the union of all local databases. An association rule* r *in* D *is heavy if* $\text{supp}_s(\text{r, D}) \geq \mu$, *and* $\text{conf}_s(\text{r, D}) \geq \nu$, *where* μ *and* ν *are the user-defined thresholds of high-support and high-confidence used for identifying heavy association rules in* D, *respectively.*

Apart from synthesized support and synthesized confidence of an association rule, the frequency of an association rule is an important issue in multi-database mining. We define *frequency* of an association rule as the number of extractions of the association rule from different databases. If an association rule is extracted from k out of n databases then the frequency of the association rule is k, for $0 \leq k \leq n$. An association rule may be high-frequency rule or, low-frequency rule, or neither high-frequency rule nor low-frequency rule in multiple databases. We could arrive in such a conclusion only if we have user-defined thresholds of low-frequency (γ_1) and high-frequency (γ_2) of an association rule, for $0 < \gamma_1 < \gamma_2 \leq 1$. A low-frequency association rule is extracted from less than $n \times \gamma_1$ databases. On the other hand, a high-frequency association rule is extracted from at least $n \times \gamma_2$ databases. In the context of multi-database mining using local pattern analysis, we define a high-frequency association rule and a low-frequency association rule as follows:

Definition 2.3 *Let an association rule be extracted from* k *out of* n *databases. Then the association rule is low-frequency rule if* $\text{k} < \text{n} \times \gamma_1$, *where* γ_1 *is the user-defined threshold of low-frequency.*

Definition 2.4 *Let an association rule be extracted from* k *out of* n *databases. Then the association rule is high-frequency rule if* $\text{k} \geq \text{n} \times \gamma_2$, *where* γ_2 *is the user-defined threshold of high-frequency.*

While synthesizing heavy association rules in multiple databases, it may be worth noting some other attributes of a synthesized association rule. For example, high-frequency, low-frequency, and exceptionality are interesting as well as important attributes of a synthesized association rule. We have already defined high-frequency association rule and low-frequency association rule in multiple databases. We now define an exceptional association rule in multiple databases as follows:

Definition 2.5 *A heavy association rule in multiple databases is exceptional if it is a low-frequency rule.*

It may be worth contrasting between a heavy association rule, a high-frequency association rule and an exceptional association rule in multiple databases.

- An exceptional association rule is also a heavy association rule.
- A high-frequency association rule is not an exceptional association rule, and vice versa.
- A high-frequency association rule is not necessarily be a heavy association rule.
- There may exist heavy association rules that are neither high-frequency rule nor exceptional rule.

The goal of this chapter is to present an extended model of local pattern analysis. Also, we see later how this model helps mining extreme types of association rules as identified above.

2.3 An Extended Model of Local Pattern Analysis for Synthesizing Global Patterns from Local Patterns in Different Databases

Let D_i be the database corresponding to i-th branch of the organization, $i = 1, 2, \ldots, n$. Patterns in multiple databases could be grouped into following categories based on the number of databases: local patterns, global patterns, and patterns that are neither local nor global. A pattern based on a branch database is called a *local pattern*. On the other hand, a *global pattern* is based on all databases under consideration. An essence of the extended model of local pattern analysis (Adhikari and Rao 2008) is illustrated in Fig. 2.1. The extended model comes with a set of interfaces and a set of layers. Each interface realizes a set of operations that produces dataset(s) (or, knowledge) based on the dataset(s) available at the next lower layer. There are four interfaces of the proposed model of synthesizing global patterns from local patterns.

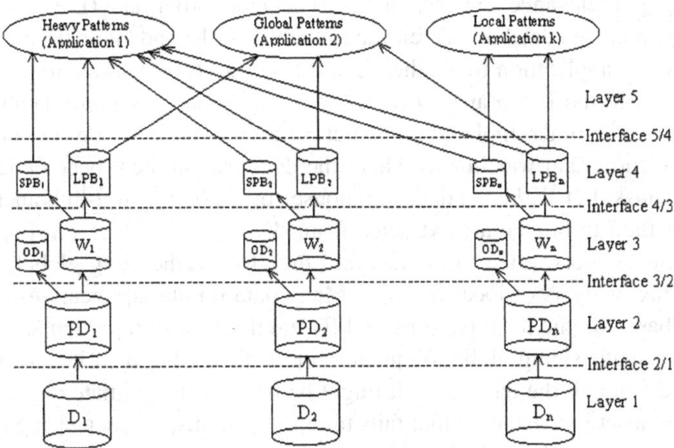

Fig. 2.1 A model of synthesizing global patterns from local patterns in different databases

Interface 2/1 is concerned with different operations on data realized at the lowest layer. By applying these operations, we come up with a processed database resulting from a local (original) database. These operations are performed on each branch database. Interface 3/2 applies a filtering algorithm to each processed database to separate relevant data from outliers. In particular, if we are interested in studying durable items then the transactions containing only non-durable items could be treated as outlier transactions. Different interesting criteria could be set to filter data.

This interface supports loading data into the respective data warehouse. Interface 4/3 mines (local) patterns in each local data warehouse. There are two types of local patterns: local patterns and suggested local patterns. A suggested local pattern is close but fails to fully satisfy the requisite interestingness criteria. The reasons for considering suggested patterns are given as follows. Firstly, by admitting these patterns, we could synthesize patterns more accurately. Secondly, due to the stochastic nature of the transactions, the number of suggested patterns could be significant in some databases. Thirdly, there is a tendency that a suggested pattern of one database could become a local pattern in some other databases. Thus, the correctness of synthesizing global patterns would increase as the number of local patterns increases. Therefore, the extended model becomes effective in synthesizing non-local patterns. Consider a multi-branch company having n databases. Let LPB_i and SPB_i be the local pattern base and suggested local pattern base corresponding to i-th branch of the organization, respectively, $i = 1, 2, \ldots, n$. Interface 5/4 synthesizes global patterns, or analyses local patterns for finding solutions to many problems.

At the lowest layer, all the local databases are kept. We may need to process these databases for the purpose of data mining task. Various data preparation techniques (Pyle 1999) – data preprocessing like data cleaning, data transformation, data integration, and data reduction are applied to data in the local databases. We get the processed database PD_i corresponding to the original database D_i, for $i = 1, 2, \ldots, n$. Then we retain all the data that are relevant to the data mining applications. Using a relevance analysis, one could detect outlier data (Last and Kandel 2001) from processed database. A relevance analysis is dependent on the context and varies from one application to another application. Let OD_i be the outlier database corresponding to the i-th branch, $i = 1, 2, \ldots, n$. Sometimes these databases are also used in some other applications. After removing outliers from the processed database we form data warehouse, where the data present there become ready for data mining task. Let W_i be the data warehouse corresponding to i-th branch. Local patterns for the i-th branch are extracted from W_i, for $i = 1, 2, \ldots, n$. Finally, the local patterns are forwarded to the central office for synthesizing global patterns, or completing analysis of local patterns. Many data mining applications could be developed based on the local patterns in different databases. In particular, if we are interested in synthesizing global frequent itemsets then a frequent itemset may not be extracted from all the databases. It might be required to estimate the support of a frequent itemset in a database that fails to report it. Thus, in essence, a global frequent itemset synthesized from local frequent itemsets is approximate. If any one of the local databases is too large to apply a traditional data mining technique then this model would fail. In this situation, one could apply an appropriate sampling technique to reduce the size of the corresponding local database. Otherwise, the database could be partitioned into sub-databases. As a result, the error of data analysis would increase.

Though the above model introduces many layers and interfaces for synthesizing global patterns, in a real life application, some of these layers might not be fully exploited. In the following section, we discuss a problem of multi-database mining that uses the above model.

2.4 An Application: Synthesizing Heavy Association Rules in Multiple Real Databases

In the previous section, we have discussed different types of extreme association rules. In this section, we present an algorithm for synthesizing heavy association rules in multiple databases. The algorithm also notifies the high-frequency and exceptionality statuses of heavy association rules.

As discussed in Chapter 1, we have observed some difficulties in extracting heavy association rules in the union of all branch databases by employing a traditional data mining technique. Therefore, we synthesize heavy association rules using patterns in branch databases. Let *D* be the union of all branch databases. Also, let RB_i and SB_i be the rulebase and suggested rulebase corresponding to database D_i, respectively. An association rule $r \in RB_i$, if $supp_a(r, D_i) \geq \alpha$, and $conf_a(r, D_i) \geq \beta$, $i = 1, 2, \ldots, n$. An association rule $r \in SB_i$, if $supp_a(r, D_i) \geq \alpha$, and $conf_a(r, D_i) < \beta$. There is a tendency of a suggested association rule in a database to become an association rule in another database. Apart from the association rules, we also consider the suggested association rules for synthesizing heavy association rules in *D*. The reasons for considering suggested association rules are given as follows. Firstly, we could synthesize support and confidence of an association rule in *D* more accurately. Secondly, we could synthesize high-frequency association rules in *D* more accurately. Thirdly, some experimental results have shown that the number of suggested association rules could be significant for some databases. In general, the accuracy of synthesizing an association rule increases as the number of extractions of the association rule increases. Thus, we consider suggested association rules also in synthesizing heavy association rules in *D*. In addition, the number of transactions in a database would be required for synthesizing an association rule. We define *size* of database *DB* as the number of transactions in *DB*, denoted by *size*(*DB*). We state the application problem as follows.

Let there are n distinct databases $D_1, D_2, \ldots, D_n$. Let RB_i and SB_i be the set of association rules and suggested association rules in D_i, respectively, $i = 1, 2, \ldots, n$. Synthesize heavy association rules in the union of all databases (D) based on RB_i and SB_i, $i = 1, 2, \ldots, n$. Also, notify whether each heavy association rule is high-frequency rule or exceptional rule in D.

2.4.1 Related Work

Some applications of multiple large databases have been discussed in Chapter 1. Association rule mining gives rise to interesting association between two itemsets in a database. The notion of association rule is introduced by Agrawal et al. (1993). The authors have proposed an algorithm to mine frequent itemsets in a database. Many algorithms to extract association rules have been reported in the literature. In what follows, we present a few interesting algorithms for extracting association rules in a database. Agrawal and Srikant (1994) have proposed apriori algorithm that uses breadth-first search strategy to count the supports of itemsets. The algorithm uses an improved candidate generation function, which exploits the downward closure property of support and makes it more efficient than earlier algorithm. Han et al. (2000) have proposed data mining method of FP-growth (frequent pattern growth) which uses an extended prefix-tree (FP-tree) structure to store the database in a compressed form. FP-growth adopts a divide-and-conquer approach to decompose both the mining tasks and databases. It uses a pattern fragment growth method to avoid the costly process of candidate generation and testing. Savasere et al. (1995) have introduced partition algorithm. The database is scanned only twice. In the first scan, the database is partitioned and in each partition support is counted. Then the counts are merged to generate potential frequent itemsets. In the second scan, the potential frequent itemsets are counted to find the actual frequent itemsets.

Existing parallel mining techniques (Agrawal and Shafer 1999; Chattratichat et al. 1997; Cheung et al. 1996) could also be used to mine heavy association rules in multi-databases. Zhong et al. (2003) have proposed a theoretical framework for peculiarity oriented mining in multiple data sources. Zhang et al. (2009) have proposed a nonlinear method, named KEMGP, which adopts kernel estimation method for synthesizing global patterns from local patterns. Shang et al. (2008) have proposed an extension to Piatetsky-Shapiro's minimum interestingness condition to mine association rules in multiple databases.

Yi and Zhang (2007) have proposed a privacy-preserving distributed association rule mining protocol based on a semi-trusted mixer model. Rozenberg and Gudes (2006) have presented their work on association rule mining from distributed vertically partitioned data with the goal of preserving the confidentiality of each database. The authors have presented two algorithms for discovering frequent itemsets and for calculating the confidence of the rules.

2.4.2 Synthesizing an Association Rule

The technique of synthesizing heavy association rules is suitable for the real databases, where the trend of the customers' behavior exhibited in one database is usually present in other databases. In particular, a frequent itemset in one database is usually present in some transactions of other databases even if it does not get extracted. Our estimation procedure captures such trend and estimates the support of a missing association rule in a database. Let $E_1(r, DB)$ be the amount of error in estimating support of a missing association rule r in database DB. Also, let $E_2(r, DB)$ be the level of error in assuming support as 0 for the missing association rule in DB. Then the value of $E_1(r, DB)$ is usually lower than $E_2(r, DB)$. The estimated support and confidence of a missing association rule usually reduce the error of synthesizing heavy association rules in different databases. We would like to estimate the support and confidence of a missing association rule rather assuming it as absent in a database. If an association rule fails to get extracted from database DB, then we assume that DB contributes some amount of support and confidence for the association rule. The support and confidence of an association rule r in database DB satisfy the following inequality:

$$0 \le supp_a(r, DB) \le conf_a(r, DB) \le 1 \tag{2.1}$$

At a given $\alpha = \alpha_0$, we observe that the confidence of an association rule r varies over the interval $[\alpha_0, 1]$ as explained in Example 2.1.

Example 2.1 *Let $\alpha = 0.333$. Assume that database D_1 contains the following transactions: {a1, b1, c1}, {a1, b1, c1}, {b2, c2}, {a2, b3, c3}, {a3, b4} and {c4}. The support and confidence of association rule r: {a1}→{b1} in D_1 are 0.333 and 1.0 (highest) respectively. Assume that database D_2 contains the following transactions: {a1, b1, c1}, {a1, b1}, {a1, c1}, {a1}, {a1, b2} and {a1, b3}. The support and confidence of r in D_2 are 0.333 and 0.333 (lowest), respectively.*

As the support of an association rule is expressed as the lower bound of its confidence, the confidence goes up as support increases. The support of an association rule is distributed over [0, 1]. If an association rule is not extracted from a database, then the support falls in $[0, \alpha)$, since the suggested association rules are also considered for synthesizing association rules. We would be interested in estimating the support of such rules. Assume that the association rule r: $\{c\}\rightarrow\{d\}$ has been extracted from m databases, $1 \le m \le n$. Without any loss of generality, we assume that the association rule r has been reported from the first m databases. We shall use the average behavior of the customers of the first m branches to estimate the average behavior of the customers in remaining branches. Let $D_{i,j}$ denote the union of databases $D_i, D_{i+1}, \ldots, D_j$, for $1 \le i \le j \le n$. Then, $supp_a(\{c, d\}, D_{1,m})$ could be viewed as the average behavior of customers of the first m branches for purchasing items c and d together at the same time. Then, $supp_a(\{c, d\}, D_{1,m})$ is obtained by the following formula:

$$supp_a(\{c,d\}, D_{1,m}) = \left(\sum_{i=1}^{m} supp_a(r, D_i) \times size(D_i)\right) \Big/ \sum_{i=1}^{m} size(D_i) \tag{2.2}$$

We could estimate the support of association rule r for each of the remaining (n-m) databases as follows:

$$supp_s(r, D_{m+1,n}) = \alpha \times supp_a(\{c,d\}, D_{1,m}) \tag{2.3}$$

The number of the transactions containing the itemset $\{c, d\}$ in D_i is $supp_a(r, D_i) \times size(D_i)$, for $i = 1, 2, \ldots, m$. The association rule r is not present in D_i, for $i = m + 1, m + 2, \ldots, n$. Then the estimated number of the transactions containing the itemset $\{c, d\}$ in D_i is $supp_s(r, D_{m+1,n}) \times size(D_i)$, for $i = m + 1, m + 2, \ldots, n$. The estimated support of association rule r in D_i is determined as follows:

$$supp_e(r, D_i) = \begin{cases} supp_a(r, D_i), \text{ for } i = 1, 2, \ldots, m \\ supp_s(r, D_{m+1,n}), \text{ for } i = m+1, m+2, \ldots, n \end{cases} \tag{2.4}$$

Then the synthesized support of association rule r in D could be obtained as follows.

$$supp_s\,(r,D) = \left(\sum_{i=1}^{n} supp_e(r,D_i) \times size(D_i)\right) \Big/ \sum_{i=1}^{n} size(D_i) \tag{2.5}$$

The confidence of the association rule r depends on the supports of the itemsets $\{c\}$ and $\{c, d\}$. The support of itemset $\{c, d\}$ has been synthesized. Now, we need to synthesize the support of itemset $\{c\}$. Without any loss of generality, let the itemset $\{c\}$ gets extracted from first p databases, for $1 \leq m \leq p \leq n$. The estimated support of frequent itemset $\{c\}$ in D_i could be obtained as follows:

$$supp_e\,(\{c\},D_i) = \begin{cases} supp_a(\{c\},D_i), \text{ for } i = 1, 2,\ldots,p \\ supp_s\left(\{c\},D_{p+1,n}\right), \text{ for } i = p+1, p+2,\ldots,n \end{cases} \tag{2.6}$$

Then the synthesized support of itemset $\{c\}$ in D is determined as follows.

$$supp_s\,(\{c\},D) = \left(\sum_{i=1}^{n} supp_e(\{c\},D_i) \times size(D_i)\right) \Big/ \sum_{i=1}^{n} size(D_i) \tag{2.7}$$

Then we compute the synthesized confidence of association rule r in D.

$$conf_s(r,D) = supp_s(r,D)/supp_s(\{c\},D) \tag{2.8}$$

2.4.2.1 Design of the Algorithm

Here we present an algorithm for synthesizing heavy association rules in D. The algorithm also indicates whether a heavy association rule is high-frequency rule or exceptional rule. Let N and M be the number of association rules and the number of suggested association rules in different local databases, respectively. The association rules and suggested association rules are kept in arrays *RB* and *SB*, respectively. An association rule could be described by following attributes: *ant, con, did, supp* and *conf*. The attributes *ant, con, did, supp* and *conf* represent antecedent, consequent, database identification, support, and confidence of a rule, respectively. An attribute x of the i-th association rule of *RB* is denoted by *RB(i).x*, for $i = 1, 2, \ldots, |RB|$. All the synthesized association rules are kept in array *SR*. Each synthesized association rule could be described by following attributes: *ant, con, did, ssupp* and *sconf*. The attributes *ssupp* and *sconf* represent synthesized support and synthesized confidence of a synthesized association rule, respectively. In the context of mining heavy association rules in D, the following additional attributes are also considered: *heavy, highFreq, lowFreq* and *except*. The attributes *heavy, highFreq, lowFreq* and *except* are used to indicate whether an association rule is a heavy rule, high-frequency rule, low-frequency rule and exceptional rule in D, respectively. An attribute y of the i-th synthesized association rule of *SR* is denoted by *SR(i).y*, for $i = 1, 2, \ldots, |SR|$.

Algorithm 2.1 Synthesize heavy association rules in *D*. Also, indicate whether a heavy association rule is high-frequency rule or exceptional rule.

procedure *Association-Rule-Synthesis* (*n*, *RB*, *SB*, μ, ν, *size*, γ_1, γ_2)

Inputs:

n: number of databases
RB: array of association rules
SB: array of suggested association rules
μ: threshold of high-support for determining heavy association rules
ν: threshold of high-confidence for determining heavy association rules
size: array of the number of transactions in different databases
γ_1: threshold of low-frequency for determining low-frequency association rules
γ_2: threshold of high-frequency for determining high-frequency association rules

Outputs:
Heavy association rules along with their high-frequency and exceptionality statuses

```
copy rules of RB and SB into array R;
sort rules of R based on attributes ant and con;
calculate total number of transactions in all the databases and store it in
totalTrans;
let nSynRules = 1;
let curPos = 1;
while ( curPos ≤ |R| ) do
  calculate number of occurrences of current rule R(curPos) and store it in
  nExtractions;
  let SR(nSynRules).highFreq = false;
  if ((nExtractions / n) ≥ γ2) then
    SR(nSynRules). highFreq = true;
  end if
  let SR(nSynRules).lowFreq = false;
  if ((nExtractions / n) < γ1) then
    SR(nSynRules).lowFreq = true;
  end if
  calculate supp_s(R(curPos), D) using formula (2.5);
  calculate conf_s(R(curPos), D) using formula (2.8);
  let SR(nSynRules).heavy = false;
  if ((supp_s(SR(nSynRules), D) ≥ μ) and (conf_s(SR(nSynRules), D) ≥ ν))
  then
    SR(nSynRules).heavy = true;
  end if
  let SR(nSynRules).except = false;
  if ((SR(nSynRules) is a low-frequency rule) and (SR(nSynRules) is a heavy
  rule)) then
    SR(nSynRules).except = true;
```

```
25:     end if
26:     update index curPos for processing the next association rule;
27:     increase index nSynRules by 1;
28:   end while
29:   for each synthesized association rule τ in SR do
30:     if τ is heavy then
31:       display τ along with its high-frequency and exceptionality statuses;
32:     end if
33:   end for
end procedure
```

The above algorithm works as follows. The association rules and suggested association rules are copied into R. All the association rules in R are sorted on the pair of attributes $\{ant, con\}$, so that the same association rule extracted from different databases remains together after sorting. Thus, it would help synthesizing a single association rule at a time. The synthesis process is realized in the while-loop shown in line 6. Based on the number of extractions of an association rule, we could determine its high-frequency and low-frequency statuses. The number of extractions of current association rule has been determined as indicated in line 7. The high-frequency status of current association rule is determined – see lines 8–11. Also, the low-frequency status of current association rule is determined (lines 12–15). We synthesize support and confidence of current association rule based on (2.5) and (2.8), respectively. Once the synthesized support and synthesized confidence have been determined, we could identify the heavy and exceptional statues of current association rule. The heavy status of current association rule is determined using the part of the procedure covered in lines 18–21. Also, the exceptional status of current association rule is determined using lines 22–25. At line 26, we determine the next association rule in R for the synthesizing process. Heavy association rules are displayed along with their high-frequency and exceptionality statuses using lines 29–33.

Theorem 2.1 *The time complexity of procedure* Association-Rule-Synthesis *is* maximum $\{O((M+N) \times \log(M+N)), O(n \times (M+N))\}$, *where* N and M *are the number of association rules and the number of suggested association rules extracted from n databases.*

Proof The lines 1 and 2 take time in $O(M + N)$ and $O((M + N) \times \log(M + N))$ respectively, since there are $M + N$ rules in different local databases. The while-loop at line 6 repeats maximum $M + N$ times. Line 7 takes $O(n)$ time, since each rule is extracted maximum n number of times. Lines 8–15 take $O(1)$ time. Using formula (2.3), we could calculate the average behavior of customers of the first m databases in $O(m)$ time. Each of lines 16 and 17 takes $O(n)$ time. Lines 18–25 take $O(1)$ time. Line 26 could be executed during execution of line 7. Thus, the time complexity of while-loop 6–28 is $O(n \times (M + N))$. The time complexity of lines 29–33 is $O(M + N)$, since the number of synthesized association rules is less than or equal to $M + N$. Thus, time complexity of procedure *Association-Rule-Synthesis* is *maximum* $\{O((M + N) \times \log(M + N)), O(n \times (M + N)), O(M + N)\}$ = *maximum* $\{O((M + N) \times \log(M + N)), O(n \times (M + N))\}$.

Wu and Zhang (2003) have proposed *RuleSynthesizing* algorithm for synthesizing high-frequency association rules in different databases. The algorithm is based on the weights of the different databases. Again, the weight of a database would depend on the association rules extracted from the database. The proposed algorithm executes in $O(n^4 \times maxNosRules \times totalRules^2)$ time, where *n*, *maxNosRules*, and *totalRules* are the number of data sources, the maximum among the numbers of association rules extracted from different databases, and the total number of association rules in different databases, respectively. Ramkumar and Srinivasan (2008) have proposed a modification of *RuleSynthesizing* algorithm. In this modified algorithm, the weight of an association rule is based on the size of a database. This assumption seems to be more logical. For synthesizing confidence of an association rule, the authors have described a method which was originally proposed by Adhikari and Rao (2008). Though the time complexity of modified *RuleSynthesizing* algorithm is the same as that of original *RuleSynthesizing* algorithm, but it reduces the average error in synthesizing an association rule. The algorithm *Association-Rule-Synthesis* could synthesize heavy association rules, high-frequency association rules, and exceptional association rules in *maximum* $\{O(totalRules \times \log(totalRules)), O(n \times totalRules)\}$ time. Thus, algorithm *Association-Rule-Synthesis* takes much less time than the existing algorithms. Moreover, the proposed algorithm is simple and straight forward. We illustrate the proposed algorithm using the following example.

Example 2.2 *Let D_1, D_2 and D_3 be three databases of sizes 4,000 transactions, 3,290 transactions, and 10,200 transactions, respectively. Let D be the union of the databases D_1, D_2, and D_3. Assume that $\alpha = 0.2$, $\beta = 0.3$, $\gamma_1 = 0.4$, $\gamma_2 = 0.7$, $\mu = 0.3$ and $\nu = 0.4$. The following association rules have been extracted from the given databases. r_1: $\{H\} \rightarrow \{C, G\}$, r_2: $\{C\} \rightarrow \{G\}$, r_3: $\{G\} \rightarrow \{F\}$, r_4: $\{H\} \rightarrow \{E\}$, r_5: $\{A\} \rightarrow \{B\}$. The rulebases are given as follows: $RB_1 = \{r_1, r_2\}$, $SB_1 = \{r_3\}$; $RB_2 = \{r_4\}$, $SB_2 = \{r_1\}$; $RB_3 = \{r_1, r_5\}$, $SB_3 = \{r_2\}$. The supports and confidences of the association rules are given as follows. $supp_a(r_1, D_1) = 0.22$, $conf_a(r_1, D_1) = 0.55$; $supp_a(r_1, D_2) = 0.25$, $conf_a(r_1, D_2) = 0.29$; $supp_a(r_1, D_3) = 0.20$, $conf_a(r_1, D_3) = 0.52$; $supp_a(r_2, D_1) = 0.69$, $conf_a(r_2, D_1) = 0.82$; $supp_a(r_2, D_3) = 0.23$, $conf_a(r_2, D_3) = 0.28$; $supp_a(r_3, D_1) = 0.22$, $conf_a(r_3, D_1) = 0.29$; $supp_a(r_4, D_2) = 0.40$, $conf_a(r_4, D_2) = 0.45$; $supp_a(r_5, D_3) = 0.86$, $conf_a(r_5, D_3) = 0.92$. Also, let $supp_a(\{A\}, D_3) = 0.90$, $supp_a(\{C\}, D_1) = 0.80$, $supp_a(\{C\}, D_3) = 0.40$, $supp_a(\{G\}, D_1) = 0.29$, $supp_a(\{H\}, D_1) = 0.31$, $supp_a(\{H\}, D_2) = 0.33$, and $supp_a(\{H\}, D_3) = 0.50$. Heavy association rules are presented in Table 2.1.*

The association rules r_2 and r_5 have synthesized support greater than or equal to 0.3 and synthesized confidence greater than or equal to 0.4. So, r_2 and r_5 are heavy association rules in D. The association rule r_5 is a exceptional rule, since

Table 2.1 Heavy association rules in the union of databases given in Example 2.2

r: ant→ con	*ant*	*con*	$supp_s(r, D)$	$conf_s(r, D)$	Heavy	High freq	Except
r_2	C	G	0.31	0.66	True	False	False
r_5	A	B	0.57	0.90	True	False	True

it is a heavy and low-frequency rule. But the association rule r_2 is neither a high-frequency nor exceptional rule. Though the association rule r_1 is a high-frequency rule but it is not a heavy rule, since $supp_s(r_1, D) = 0.21$ and $conf_s(r_1, D) = 0.48$.

2.4.3 Error Calculation

To evaluate the proposed technique of synthesizing heavy association rules we have determined the error which has occurred in the experiments. More specifically, the error is expressed relative to the number of transactions, number of items, and the length of a transaction in the databases. Thus the error of an experiment needs to be expressed along with *ANT*, *ALT*, and *ANI* in the given databases, where *ANT*, *ALT* and *ANI* denote the average number of transactions, the average length of a transaction and the average number of items in a database, respectively. There are several ways one could define the error. The proposed definition of error is based on the frequent itemsets generated from heavy association rules. Let r: $\{c\} \rightarrow \{d\}$ be a heavy association rule. Then the frequent itemsets generated from association rule r are $\{c\}$, $\{d\}$, and $\{c, d\}$. Let $\{X_1, X_2, \ldots, X_m\}$ be set of frequent itemsets generated from all the heavy association rules in D. We define the following two types of error.

1. *Average Error* (AE)

$$\text{AE}(D, \alpha, \mu, \nu) = \frac{1}{m} \sum_{i=1}^{m} |supp_a(X_i, D) - supp_s(X_i, D)| \tag{2.9}$$

2. *Maximum Error* (ME)

$$\text{ME}(D, \alpha, \mu, \nu) = maximum\{\, |supp_a(X_i, D) - supp_s(X_i, D)|\, , i = 1, 2, \ldots, m\} \tag{2.10}$$

where $supp_a(X_i, D)$ and $supp_s(X_i, D)$ are actual support i.e., the support based on apriori algorithm and synthesized support of the itemset X_i in D, respectively. In Example 2.3, we illustrate the behaviour of the measures given above.

Example 2.3 *With reference to Example 2.2, r_2: C → G and r_5: A → B are heavy association rules in D. The frequent itemsets generated from r_2 and r_5 are A, B, C, G, AB and CG. For the purpose of finding the error of an experiment, we need to find the actual support of the itemsets generated from the heavy association rules.*

The actual support of an itemset generated from a heavy association rule could be obtained by mining all the databases D_1, D_2, and D_3 together.

Thus, $\mathrm{AE}(D, 0.2, 0.3, 0.4) = \frac{1}{6}\{\, |supp_a(\{A\},D) - supp_s(\{A\},D)| +$
$|supp_a(\{B\},D) - supp_s(\{B\},D)| + |supp_a(\{C\},D) - supp_s(\{C\},D)| +$
$|supp_a(\{G\},D) - supp_s(\{G\},D)| + |supp_a(\{A,B\},D) - supp_s(\{A,B\},D)| +$
$|supp_a(\{C,G\},D) - supp_s(\{C,G\},D)|\,\}$.

$\mathrm{ME}(D, 0.2, 0.3, 0.4) = maximum\{\, |supp_a(\{A\},D) - supp_s(\{A\},D)|$,
$|supp_a(\{B\},D) - supp_s(\{B\},D)|$, $|supp_a(\{C\},D) - supp_s(\{C\},D)|$,
$|supp_a(\{G\},D) - supp_s(\{G\},D)|$, $|supp_a(\{A,B\},D) - supp_s(\{A,B\},D)|$,
$|supp_a(\{C,G\},D) - supp_s(\{C,G\},D)|\,\}$.

2.4.4 Experiments

We have carried out several experiments to study the effectiveness of the approach presented in this chapter. We present the experimental results using three real databases. The database *retail* (Frequent itemset mining dataset repository 2004) is obtained from an anonymous Belgian retail supermarket store. The databases *BMS-Web-Wiew*-1 and *BMS-Web-Wiew*-2 can be found from KDD CUP 2000 (Frequent itemset mining dataset repository 2004). We present some characteristics of these databases in Table 2.2. We use notation *DB*, *NT*, *AFI*, *ALT* and *NI* to denote a database, the number of transactions, the average frequency of an item, the average length of a transaction and the number of items in the database, respectively.

Table 2.2 Dataset characteristics

Dataset	*NT*	*ALT*	*AFI*	*NI*
retail	88,162	11.31	99.67	10,000
BMS-Web-Wiew-1	1,49,639	2.00	155.71	1,922
BMS-Web-Wiew-2	3,58,278	2.00	7,165.56	100

Each of the above databases is divided into 10 subsets for the purpose of carrying out experiments. The databases obtained from *retail*, *BMS-Web-Wiew*-1 and *BMS-Web-Wiew*-2 are named as R_i, B_{1i} and B_{2i} respectively, i = 0, 1, ..., 9. The databases R_j and B_{ij} are called branch databases, i = 1, 2, and j = 0, 1, ..., 9. Some characteristics of these branch databases are presented in Table 2.3.

Table 2.3 Branch database characteristics

DB	*NT*	*ALT*	*AFI*	*NI*	*DB*	*NT*	*ALT*	*AFI*	*NI*
R_0	9,000	11.24	12.07	8,384	R_5	9,000	10.86	16.71	5,847
R_1	9,000	11.21	12.27	8,225	R_6	9,000	11.20	17.42	5,788
R_2	9,000	11.34	14.60	6,990	R_7	9,000	11.16	17.35	5,788
R_3	9,000	11.49	16.66	6,206	R_8	9,000	12.00	18.69	5,777
R_4	9,000	10.96	16.04	6,148	R_9	7,162	11.69	15.35	5,456
B_{10}	14,000	2.00	14.94	1,874	B_{15}	14,000	2.00	280.00	100
B_{11}	14,000	2.00	280.00	100	B_{16}	14,000	2.00	280.00	100
B_{12}	14,000	2.00	280.00	100	B_{17}	14,000	2.00	280.00	100
B_{13}	14,000	2.00	280.00	100	B_{18}	14,000	2.00	280.00	100
B_{14}	14,000	2.00	280.00	100	B_{19}	23,639	2.00	472.78	100
B_{20}	35,827	2.00	1,326.93	54	B_{25}	35,827	2.00	716.54	100
B_{21}	35,827	2.00	1,326.93	54	B_{26}	35,827	2.00	716.54	100
B_{22}	35,827	2.00	716.54	100	B_{27}	35,827	2.00	716.54	100
B_{23}	35,827	2.00	716.54	100	B_{28}	35,827	2.00	716.54	100
B_{24}	35,827	2.00	716.54	100	B_{29}	35,835	2.00	716.70	100

The results of the three experiments using Algorithm 2.1 are presented in Table 2.4. The choice of different parameters is an important issue. We have selected different values of α and β for different databases. But, they are kept the same for branch databases obtained from the same database. For example, α and β are the same for branch databases R_i , for $i = 0, 1, \ldots, 9$.

Table 2.4 First five heavy association rules reported from different databases (sorted in non-increasing order on synthesized support)

Data base	α	β	μ	ν	Heavy assoc rules	Syn supp	Syn conf	High freq	Exceptional
$\cup_{i=0}^{9} R_i$	0.05	0.2	0.1	0.5	{48}→{39}	0.33	0.68	Yes	No
					{39}→{48}	0.33	0.56	Yes	No
					{41}→{39}	0.13	0.63	Yes	No
					{38}→{39}	0.12	0.66	Yes	No
					{41}→{48}	0.10	0.51	Yes	No
$\cup_{i=0}^{9} B_{1i}$	0.01	0.2	0.007	0.1	{1}→{5}	0.01	0.13	No	No
					{5}→{1}	0.01	0.11	No	No
					{7}→{5}	0.01	0.12	No	No
					{5}→{7}	0.01	0.11	No	No
					{3}→{5}	0.01	0.12	No	No
$\cup_{i=0}^{9} B_{2i}$	0.006	0.01	0.01	0.1	{3}→{1}	0.02	0.14	Yes	No
					{1}→{3}	0.02	0.14	Yes	No
					{7}→{1}	0.02	0.14	Yes	No
					{1}→{7}	0.02	0.14	Yes	No
					{5}→{1}	0.02	0.14	Yes	No

After mining a branch database from a group of branch databases using a reasonably low values α and β, one could fix α and β for the purpose data mining task. If α and β are smaller, then synthesized support and synthesized confidence values are closer to their actual values. Thus, the synthesized association rules are closer to the true association rules in multiple databases.

The choice of μ and ν are context dependent. Also if μ and ν are kept fixed then some databases might not report heavy association rules, while other databases might report many heavy association rules. While generating association rule one could estimate the average synthesized support and confidence based on the generated association rules. Thus, it gives an idea of thresholds for high-support and high-confidence for synthesizing heavy association rules in different databases. Also, the choice of γ_1 and γ_2 are also context dependent. It has been found that "reasonable" values of γ_1 and γ_2 could lie in the interval [0.3, 0.4] and [0.6, 0.7], respectively. Given these findings, we have taken $\gamma_1 = 0.35$, and $\gamma_2 = 0.60$ for synthesizing heavy association rules.

The experiments conducted on the three databases have resulted in no exceptional association rule. Normally, exceptional association rules are rare. Also, we have not found any association rule which is heavy rule as well as high-frequency rule in multiple databases obtained from *BMS-Web-Wiew-1*.

In many applications, the suggested association rules are significant. While synthesizing the association rules from different databases we might need to consider the suggested association rules for the correctness of synthesizing association rules. We have observed that the number of suggested association rules in the set of databases $\{R_0, R_1, \ldots, R_9\}$ and $\{B_{10}, B_{11}, \ldots, B_{19}\}$ are significant. But, the set of databases $\{B_{20}, B_{21}, \ldots, B_{29}\}$ do not generate any suggested association rule. We present the number of association rules and the number of suggested association rules for different experiments in Table 2.5.

Table 2.5 Number of association rules and suggested association rules extracted from multiple databases

Database	α	β	Number of association rules (N)	Number of suggested association rules (M)	$M/(N+M)$
$\cup_{i=0}^{9} R_i$	0.05	0.2	821	519	0.39
$\cup_{i=0}^{9} B_{1i}$	0.01	0.2	50	96	0.66
$\cup_{i=0}^{9} B_{2i}$	0.006	0.01	792	0	0

The error of synthesizing association rules in a database is relative to the following parameters: the number of transactions, the number of items, and the length of transactions in the given databases. If the number of transactions in database increases, the error of synthesizing association rules also increases, provided other two parameters remain constant. If the lengths of transactions of a database increase, the error of synthesizing association rules is likely to increase, provided that two other parameters remain constant. Lastly, if the number of items increases, then the

Table 2.6 Error of synthesizing heavy association rules

Database	α	β	μ	ν	(*AE*, *ANT*, *ALT*, *ANI*)	(*ME*, *ANT*, *ALT*, *ANI*)
$\cup_{i=0}^{9} R_i$	0.05	0.2	0.1	0.5	(0.00, 8,816.2, 11.31, 5,882.1)	(0.00, 8,816.2, 11.31, 5,882.1)
$\cup_{i=0}^{9} B_{1i}$	0.01	0.2	0.007	0.1	(0.00, 14,963.9, 2.0, 277.4)	(0.00, 14,963.9, 2.0, 277.4)
$\cup_{i=0}^{9} B_{2i}$	0.006	0.01	0.01	0.1	(0.000118, 35,827.8, 2.0, 90.8)	(0.00, 35,827.8, 2.0, 90.8)

error of synthesizing association rules is likely to decrease, provided that two other parameters remain constant. Thus, the error needs to be reported along with the *ANT*, *ALT* and *ANI* for the given databases. The obtained results are presented in Table 2.6.

2.4.4.1 Comparison with Existing Algorithm

In this section, we make a detailed comparison among the part of the proposed algorithm that synthesizes only high-frequency association rules, *RuleSynthesizing* algorithm (Wu and Zhang 2003) and *Modified RuleSynthesizing* algorithm (Ramkumar and Srivinasan 2008). Let the part of the proposed algorithm be *High-Frequency-Rule-Synthesis* used for synthesizing (only) high-frequency association rules in different databases. We conduct experiments for comparing these algorithms. We compare these algorithms on the basis of the following two issues: average error and execution time.

Analysis of Average Error

The definitions of average error and maximum error given above and those proposed by Wu and Zhang (2003) are similar and use the same set of synthesized frequent itemsets. However the methods of synthesizing frequent itemsets for these two approaches are different. Thus, the level of error incurred in these two approaches might differ. In *RuleSynthesizing* algorithm, if an itemset fails to get extracted from a database then the support of the itemset is assumed to be 0. But in *Association-Rule-Synthesis* algorithm, if an itemset fails to get extracted from a database then the support of the itemset is estimated. The synthesized support of an itemset in the union of databases in these two approaches might be different. As the number of databases increases the relative presence of a rule normally decreases. Thus, the error of synthesizing an association rule normally increases. So the AE reported in the experiment is likely to increase if the number of databases increases. We observe such phenomenon in Figs. 2.2 and 2.3.

The proposed algorithm follows a direct approach in identifying high-frequency association rules as opposed to the *RuleSynthesizing* and *Modified RuleSynthesizing* algorithms. In Figs. 2.2 and 2.3, we observe that AE of an experiment conducted using *High-Frequency-Rule-Synthesis* algorithm is less than that of

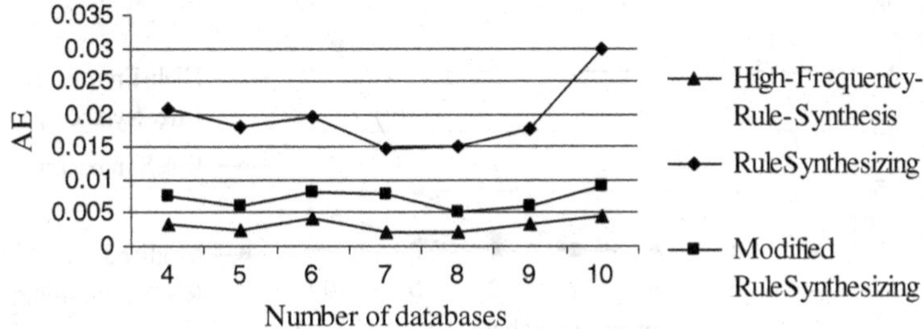

Fig. 2.2 AE vs. number of databases from *retail* at $(\alpha, \beta, \gamma) = (0.05, 0.2, 0.6)$

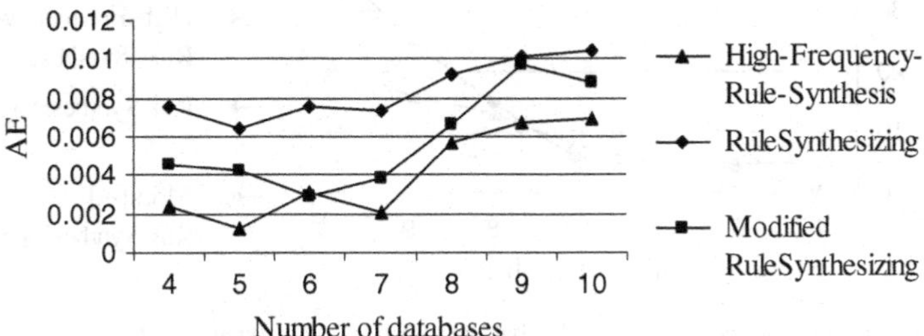

Fig. 2.3 AE vs. number of databases from *BMS-Web-Wiew*-1 at $(\alpha, \beta, \gamma) = (0.005, 0.1, 0.3)$

RuleSynthesizing algorithm. But the *Modified RuleSynthesizing* algorithm improves the accuracy of synthesizing an association rule as compared to *RuleSynthesizing* algorithm. It remains less accurate when compared to the *High-Frequency-Rule-Synthesis* algorithm.

Analysis of Execution Time

We have also completed experiments to study the execution time by varying the number of databases. The number of synthesized frequent itemsets increases as the number of databases increases. The execution time increases with the increase of number of databases. We observe this phenomenon in Figs. 2.4 and 2.5. However, more significant differences are noted with the increase in the number of databases.

The time complexities of *RuleSynthesizing* and *Modified RuleSynthesizing* algorithms are the same. When the number of databases is less the *RuleSynthesizing* and *Modified RuleSynthesizing* algorithms might be faster than *High-Frequency-Rule-Synthesizing* algorithm. As the number of databases increases, *High-Frequency-Rule-Synthesizing* algorithm works faster than both *RuleSynthesizing* and *Modified RuleSynthesizing* algorithms.

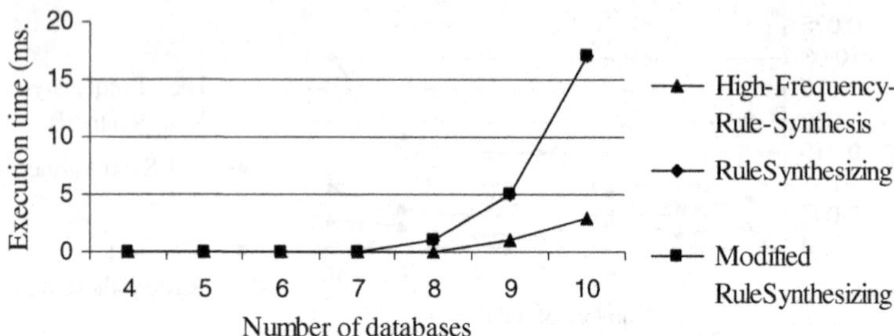

Fig. 2.4 Execution time vs. number of databases from *retail* at $(\alpha, \beta, \gamma) = (0.05, 0.2, 0.6)$

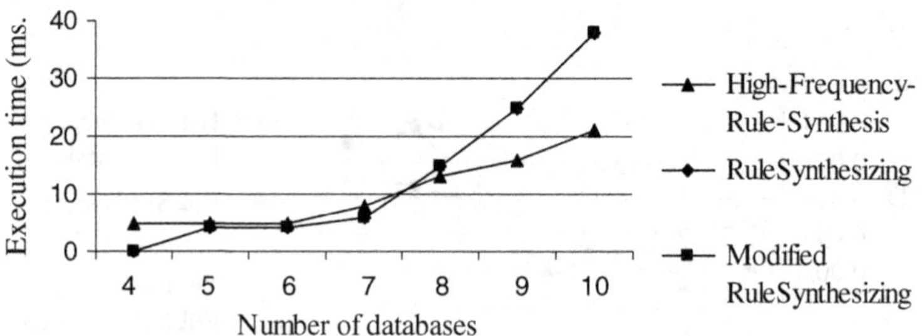

Fig. 2.5 Execution time vs. number of databases from *BMS-Web-Wiew-1* at $(\alpha, \beta, \gamma) = (0.005, 0.1, 0.3)$

2.5 Conclusions

The extended model of local pattern analysis enables us to develop useful multi-database mining applications. Although it exhibits many layers and interfaces, this general model can come with many variations. In particular, some of these layers might not be present when developing a particular application. Synthesizing heavy association rule is an important component of a multi-database mining system. In this chapter, we have presented three extreme types of association rules present in multiple databases viz., heavy association rules, high-frequency association rules and exceptional association rules. The introduced algorithm referred to as the *Association-Rule-Synthesis* is used to synthesize these extreme association rules in multiple databases.

References

Adhikari A, Rao PR (2008) Synthesizing heavy association rules from different real data sources. Pattern Recognition Letters 29(1): 59–71

Agrawal R, Imielinski T, Swami A (1993) Mining association rules between sets of items in large databases. In: Proceedings of ACM SIGMOD Conference, Washington, DC, pp. 207–216

Agrawal R, Shafer J (1999) Parallel mining of association rules. IEEE Transactions on Knowledge and Data Engineering 8(6): 962–969

Agrawal R, Srikant R (1994) Fast algorithms for mining association rules. In: Proceedings of International Conference on Very Large Data Bases, pp. 487–499

Chattratichat J, Darlington J, Ghanem M, Guo Y, Hüning H, Köhler M, Sutiwaraphun J, To HW, Yang D (1997) Large scale data mining: Challenges, and responses. In: Proceedings of the Third International Conference on Knowledge Discovery and Data Mining, pp. 143–146

Cheung D, Ng V, Fu A, Fu Y (1996) Efficient mining of association rules in distributed databases. IEEE Transactions on Knowledge and Data Engineering 8(6): 911–922

Frequent itemset mining dataset repository (2004) http://fimi.cs.helsinki.fi/data

Han J, Pei J, Yiwen Y (2000) Mining frequent patterns without candidate generation. In: Proceedings of ACM SIGMOD Conference on Management of Data, Dallas, TX, pp. 1–12

Last M, Kandel A (2001) Automated detection of outliers in real-world data. In: Proceedings of the Second International Conference on Intelligent Technologies, Bangkok, pp. 292–301

Pyle D (1999) Data Preparation for Data Mining. Morgan Kufmann, San Francisco

Ramkumar T, Srivinasan R (2008) Modified algorithms for synthesizing high-frequency rules from different data sources. Knowledge and Information Systems 17(3): 313–334

Rozenberg B, Gudes E (2006) Association rules mining in vertically partitioned databases. Data and Knowledge Engineering 59(2): 378–396

Savasere A, Omiecinski E, Navathe S (1995) An efficient algorithm for mining association rules in large databases. In: Proceedings of the 21st International Conference on Very Large Data Bases, Zurich, Switzerland, pp. 432–443

Shang S, Dong X, Li J, Zhao Y (2008) Mining positive and negative association rules in multi-database based on minimum interestingness. In: Proceedings of the 2008 International Conference on Intelligent Computation Technology and Automation 01, Washington, DC, pp. 791–794

Wu X, Zhang S (2003) Synthesizing high-frequency rules from different data sources. IEEE Transactions on Knowledge and Data Engineering 14(2): 353–367

Yi X, Zhang Y (2007) Privacy-preserving distributed association rule mining via semi-trusted mixer. Data and Knowledge Engineering 63(2): 550–567

Zhang S, Wu X, Zhang C (2003) Multi-database mining. IEEE Computational Intelligence Bulletin 2(1): 5–13

Zhang S, You X, Jin Z, Wu X (2009) Mining globally interesting patterns from multiple databases using kernel estimation. Expert Systems with Applications: An International Journal 36(8): 10863–10869

Zhong N, Yao YYY, Ohishima M (2003) Peculiarity oriented multidatabase mining. IEEE Transactions on Knowledge and Data Engineering 15(4): 952–960

Chapter 3
Mining Multiple Large Databases

Effective data analysis using multiple databases requires highly accurate patterns. As the local pattern analysis might extract patterns of low quality from multiple databases, it becomes necessary to improve mining multiple databases. In this chapter, we present an idea of multi-database mining by making use of local pattern analysis. We elaborate on the existing specialized and generalized techniques which are used for mining multiple large databases. In the sequel, we discuss a certain generalized technique, referred to as a pipelined feedback model, which is of particular relevance for mining multiple large databases. It significantly improves the quality of the synthesized global patterns. We define two types of error occurring in multi-database mining techniques. Experimental results are provided and they are reported for both real-world and synthetic databases. They help us assess the effectiveness of the pipelined feedback model.

3.1 Introduction

As underlined earlier, many large companies operate from a number of branches usually located at different geographical regions. Each branch collects data continuously and local data become stored locally. The collection of all branch databases might be large. Many corporate decisions of a multi-branch company are based on data stored over the branches. The challenges are to make meaningful decisions which are based on large volume of distributed data. This creates not only risk but also offers opportunities. One of the risks is a significant amount investment on hardware and software to deal with multiple large databases. The use of inefficient data mining techniques has to be taken into account and in many scenarios this shortcoming could be very detrimental to the quality of results.

Based on the number of data sources, patterns in multiple databases could be classified into three categories. These are local patterns, global patterns and patterns that are neither local nor global. A pattern based on a single database is called a local pattern. Local patterns are useful for local data analysis, and locally restricted decision making activities (Adhikari and Rao 2008b; Wu et al. 2005; Zhang et al. 2004b). On the other hand, global patterns are based on all the databases taken

A. Adhikari et al., *Developing Multi-database Mining Applications*, Advanced Information and Knowledge Processing, DOI 10.1007/978-1-84996-044-1_3,

into consideration. They are useful for data analyses of global nature (Adhikari and Rao 2008a; Wu and Zhang 2003) and global decision making problems. The intent of this chapter is to introduce and analyze a certain global model of data mining, referred to as a pipelined feedback model (PFM) (Adhikari et al. 2007b) which is used for mining/synthesizing global patterns in multiple large databases.

In Section 3.2, we formalize the idea of multi-database mining using local pattern analysis. Next, we discuss existing generalized multi-database mining techniques (Section 3.3). We analyze the existing specialized multi-database mining techniques in Section 3.4. The pipelined feedback model for mining multiple large databases is covered in Section 3.5. We also define a way in which an error associated with the model is quantified (Section 3.6). In Section 3.7, we provide experimental results using both synthetic and real-world databases.

3.2 Multi-database Mining Using Local Pattern Analysis

Consider a large company that deals with multiple large databases. For mining multiple databases, we are faced with three scenarios viz., (i) Each of the local databases is small, so that a single database mining technique (SDMT) could mine the union of all databases. (ii) At least one of the local databases is large, so that a SDMT could mine every local database, but fail to mine the union of all local databases. (iii) At least one of the local databases is very large, so that a SDMT fails to mine at least one local database. We are faced with challenges when handling the cases (ii) and (iii) and these challenges are inherently present because of the large size of some databases.

The first question which comes to our mind is whether a traditional data mining technique (Agrawal and Srikant 1994; Han et al. 2000; Coenen et al. 2004) could provide a sound solution when dealing with multiple large databases. To apply a "traditional" data mining technique we need to amass all the branch databases together. In such cases, a traditional data mining technique might not offer a good solution due to the following reasons:

- It might not be suitable as it requires heavy investment on hardware and software to deal with a large volume of data.
- A single computer might take unreasonable amount of time to mine a huge amount of data.
- It is difficult to identify local patterns if a traditional data mining technique is applied to the collection of all local databases.

In light of these problems and associated constraints, as encountered so far there have been attempts to deal with multi-database mining in a different way. Zhang et al. (2003) designed a multi-database mining technique (MDMT) using local pattern analysis. Multi-database mining using local pattern analysis could be classified into two categories viz., the techniques that analyze local patterns, and the techniques that analyze approximate local patterns. A multi-database mining technique using local pattern analysis could be viewed as a two-step process, denoted symbolically as M+S. Its essence can be explained as follows:

- Mine each local database using a SDMT by applying the model M (Step 1)
- Synthesize patterns using the algorithm S (Step 2)

We use the notation of M+S to stress a character of a multi-database mining technique in which we first use the model of mining (M) being followed by the synthesizing algorithm S.

One could apply sampling techniques (Babcock et al. 2003) for taming large volume of data. If an itemset is frequent in a large dataset then it is likely to be frequent in a sample dataset. Thus, one can mine patterns approximately in a large dataset by analyzing patterns in a representative sample dataset.

In addition to generalized multi-database mining techniques, there exist also specialized multi-database mining techniques. In what follows, we discuss some of the existing multi-database mining techniques.

3.3 Generalized Multi-database Mining Techniques

There is a significant variety of techniques that can be used in the multi-database mining applications.

3.3.1 Local Pattern Analysis

Under this model of mining multiple databases, each branch requires to mine the database using a traditional data mining technique. Afterwards, each branch is required to forward the pattern base to the central office. Then the central office processes the locally processed pattern bases collected from different branches to synthesize the global patterns and subsequently to support decision-making activities. Zhang et al. (2003) designed a multi-database mining technique (MDMT) using local pattern analysis. In Chapter 1, we presented this model in detail. We have proposed an extended model of local pattern analysis (Adhikari and Rao 2008a). It improves the quality of synthesized global patterns in multiple databases. In addition, it supports a systematic approach to synthesize the global patterns. In Chapter 2, we have presented the extended model of local pattern analysis for mining multiple large databases.

3.3.2 **Partition** *Algorithm*

For the purpose of mining multiple databases, one could apply *partition algorithm* (PA) proposed by Savasere et al. (1995). In Chapter 1, we have presented this model.

3.3.3 IdentifyExPattern *Algorithm*

Zhang et al. (2004a) have proposed algorithm, *IdentifyExPattern* (IEP) for identifying global exceptional patterns in multi-databases. Every local database is mined separately at *random order* (RO) using a SDMT to synthesize global exceptional patterns. For identifying global exceptional patterns in multiple databases, the following pattern synthesizing approach has been proposed. A pattern in a local database is assumed as absent, if it does not get reported. Let $supp_a(p, DB)$ and $supp_s(p, DB)$ be the actual (i.e., apriori) support and synthesized support of pattern p in database DB. Let D be the union of all local databases. Then support of pattern p has been synthesized in D based on the following expression:

$$supp_s(p, D) = \frac{1}{num(p)} \sum\nolimits_{i=1}^{num(p)} (supp_a(p, D_i) - \alpha) \big/ (1-\alpha) \tag{3.1}$$

where $num(p)$ is the number of databases that report p at a given minimum support level (α).

The size (i.e., the number of transactions) of a local database and support of an itemset in a local database seem to be important parameters that are used to determine the presence of an itemset in a database, since the number of transactions containing the itemset X in a database D_1 is equal to $supp(X, D_1) \times size(D_1)$. The major concern in this investigation is that the algorithm does not consider the size of a local database to synthesize the global support of a pattern. Using the IEP algorithm, the global support of a pattern has been synthesized using only supports of the pattern present in local databases.

3.3.4 RuleSynthesizing *Algorithm*

Wu and Zhang (2003) have proposed *RuleSynthesizing* (RS) algorithm for synthesizing high-frequency association rules in multiple databases. Using this technique, every local database is mined separately at *random order* (RO) using a SDMT for synthesizing high-frequency association rules. A pattern in a local database is assumed as absent, if it does not get reported. Based on the association rules present in different databases, the authors have estimated weights of different databases. Let w_i be the weight of the i-th database, $i = 1, 2, \ldots, n$. Without any loss of generality, let the association rule r be extracted from first m databases, for $1 \leq m \leq n$. Here, $supp_a(r, D_i)$ has been assumed as 0, for $i = m + 1, m + 2, \ldots, n$. Then the support of r in D has been determined in the following way:

$$supp_s(r, D) = w_1 \times supp_a(r, D_1) + \ldots + w_m \times supp_a(r, D_m) \tag{3.2}$$

Algorithm *RuleSynthesizing* offers an indirect approach for synthesizing association rules in multiple databases. Thus the time complexity of the algorithm is reasonably high. The algorithm executes in $O(n^4 \times maxNosRules \times totalRules^2)$ time, where n, *maxNosRules*, and *totalRules* are the number of data sources, the maximum among the numbers of association rules extracted from different databases, and the total number of association rules in different databases, respectively.

3.4 Specialized Multi-database Mining Techniques

For finding solution to a specific application, it might be possible to devise a better multi-database mining technique. In this section, we elaborate in detail on three specific multi-database mining techniques.

3.4.1 Mining Multiple Real Databases

We have proposed algorithm *Association-Rule-Synthesis* (ARS) for synthesizing association rules in multiple real data sources (Adhikari and Rao 2008a). The algorithm uses the model shown in Fig. 2.1. While synthesizing an association rule, it uses a specific method which is explained as follows: For real databases, the trend of the customers' behaviour exhibited in a single database is usually present in other databases. In particular, a frequent itemset in one database is usually present in some transactions of other databases even if it does not get extracted. The proposed estimation procedure captures such trend and estimates the support of a missing association rule. Without any loss of generality, let the itemset X be extracted from first m databases, for $1 \leq m \leq n$. Then trend of X in first m databases could be expressed as follows:

$$trend^{1,m}(X|\alpha) = \frac{1}{\sum_{i=1}^{m}|D_i|} \times \sum_{i=1}^{m} (supp_{\mathrm{a}}(X, D_i) \times |D_i|) \tag{3.3}$$

The number of transactions in a database could be considered as its weight. In (3.3), the trend of X in first m databases is estimated as a weighted sum of supports in the first m databases. We can use the detected trend of X encountered in the first m databases for synthesizing support of X in D. We estimate the support of X in database D_j by computing the expression $\alpha \times trend^{1,m}(X \mid \alpha), j = k+1, k+2, \ldots, n$. Then the synthesized support of X is determined as follows:

$$supp_{\mathrm{s}}(\mathrm{X}, \mathrm{D}) = \frac{trend^{1,m}(X|\alpha)}{\sum_{i=1}^{n}|D_i|} \times \left[(1-\alpha) \times \sum_{i=1}^{m}|D_i| + \alpha \times \sum_{i=1}^{n}|D_i|\right] \tag{3.4}$$

Association-Rule-Synthesis algorithm might return approximate global patterns, since an itemset might not get extracted from all the databases.

3.4.2 Mining Multiple Databases for the Purpose of Studying a Set of Items

Many important decisions are based on a set of specific items called the *select items*. We have proposed a technique for mining patterns of select items in multiple databases (Adhikari and Rao 2007a).

3.4.3 Study of Temporal Patterns in Multiple Databases

Adhikari and Rao (2009) have proposed a technique for clustering items in multiple databases based on their level of stability where a certain stability measure is used to quantify this feature. Web sites and transactional databases contain a large amount of time-stamped data related to an organization's suppliers and/or customers activities that have been reported over time. Mining these time-stamped data could help business leaders make better decisions by listening to their suppliers or customers via their transactions collected over time. Taking advantage of the model visualized in Fig. 3.1, we can extract global patterns in multiple temporal databases.

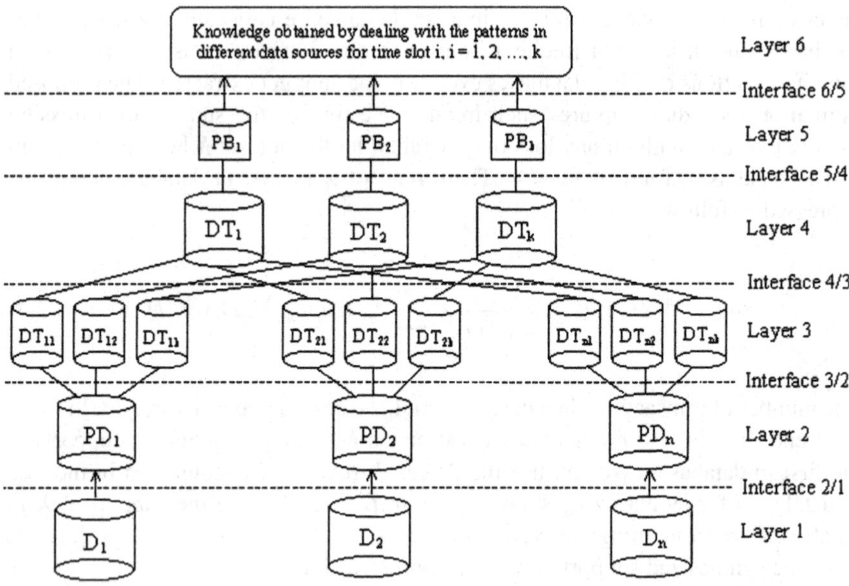

Fig. 3.1 A model of mining global patterns in multiple time-stamped databases

After a careful analysis, we note that the model shown there exhibits some commonalities with the model we showed in Fig. 2.1. Here we underline the most visible differences between the models. The model comes with a set of interfaces and is structured in a series of layers. Each interface is a set of operations that produces dataset(s) (or, knowledge) based on the lower layer dataset(s). At interface labeled 3/2, the processed database PD_i is partitioned into k time databases DT_{ij}, where DT_{ij}

is the processed database (if available) for the j-th time slot at the i-th branch, $j = 1, 2, \ldots, k$, and $i = 1, 2, \ldots, n$. The j-th time databases of all the branches are merged into a single time database $DT_j, j = 1, 2, \ldots, k$. A traditional data mining technique is applied to DT_j at the interface 5/4, for $j = 1, 2, \ldots, k$. Let PB_j be pattern base corresponding to the time database $DT_j, j = 1, 2, \ldots, k$. Finally, all the pattern bases are processed to synthesize knowledge or, take some decisions at the interface 6/5. Other lines in Fig. 3.1 are assumed to be directed from bottom to top.

Layer 4 contains all the time databases. If any one of these databases is too large to apply a traditional data mining technique then the model would fail. In this situation, one can apply an appropriate sampling technique to reduce the size of a database. In this case, one can obtain approximate patterns in databases over time.

3.5 Mining Multiple Databases Using Pipelined Feedback Model (PFM)

Before applying the pipelined feedback model, one needs to prepare data warehouses at different branches of a multi-branch organization. In Fig. 2.1, we have shown how to preprocess data warehouse at each branch. Let W_i be the data warehouse corresponding to the i-th branch, $i = 1, 2, \ldots, n$. Then the local patterns for the i-th branch are extracted from $W_i, i = 1, 2, \ldots, n$. We mine each data warehouse using any SDMT technique. In Fig. 3.2, we present a model of mining multiple databases (Adhikari et al. 2007b).

In PFM, W_1 is mined using a SDMT and as result a local pattern base LPB_1 becomes extracted. While mining W_2, all the patterns in LPB_1 are extracted irrespective of their values of interestingness measures like, minimum support and minimum confidence. Apart from these patterns, some new patterns that satisfy user-defined threshold values of interestingness measures are also extracted. In general, while mining W_i, all the patterns in W_{i-1} are mined irrespective of their values of interestingness measures, and some new patterns that satisfy user-defined threshold values of interestingness measures, $i = 2, 3, \ldots, n$. Due to this nature of mining each data warehouse, the technique is called a feedback model. Thus, $|LPB_{i-1}| \leq |LPB_i|$, for $i = 2, 3, \ldots, n$. There are $n!$ arrangements of pipelining for n databases. All the arrangements of data warehouses might not produce the same result of mining. If the number of local patterns increases, one gets more accurate global patterns

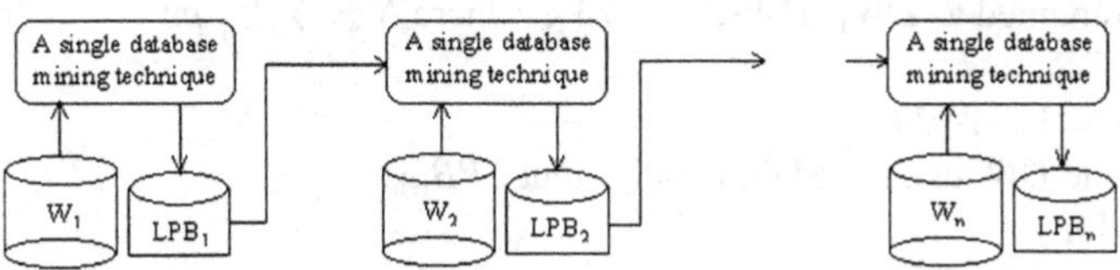

Fig. 3.2 Pipelined feedback model of mining multiple databases

which leads to a better analysis of local patterns. An arrangement of data warehouses would produce near optimal result if the cardinality $|LPB_n|$ is maximal. Let $size(W_i)$ be the size of W_i (in bytes), $i = 1, 2, \ldots, n$. We will adhere to the following rule of thumb regarding the arrangements of data warehouses for the purpose of mining. The number of patterns in W_i is greater than or equal to the number of patterns in W_{i-1} when $size(W_i) \geq size(W_{i-1})$, $i = 2, 3, \ldots, n$. For the purpose of increasing the number of local patterns, W_{i-1} precedes W_i in the pipelined arrangement of mining data warehouses if $size(W_{i-1}) \geq size(W_i)$, $i = 2, 3, \ldots, n$. Finally, we analyze the patterns in $LPB_1, LPB_2, \ldots,$ and LPB_n to synthesize global patterns, or analyze local patterns.

Let W be the collection of all branch data warehouses. For synthesizing global patterns in W we discuss here a simple pattern synthesizing (SPS) algorithm. Without any loss of generality, let the itemset X be extracted from first m databases, for $1 \leq m \leq n$. Then the synthesized support of X in W comes in the form.

$$supp_{\mathrm{s}}(X, W) = \frac{1}{\sum_{i=1}^{n} |W_i|} \times \sum_{i=1}^{m} \left[supp_{\mathrm{a}}\,(X, W_i) \times |W_i| \right] \tag{3.5}$$

3.5.1 Algorithm Design

In this section, we present an algorithm for mining multiple large databases. The method is based on the pipelined feedback model discussed above.

Algorithm 3.1 Mine multiple data warehouses using pipelined feedback model.
procedure *PipelinedFeedbackModel* ($W_1, W_2, \ldots, W_n$)
Input: $W_1, W_2, \ldots, W_n$
Output: local pattern bases

for $i = 1$ to n **do**
  if W_i does not fit in memory **then**
    partition W_i into $W_{i1}, W_{i2}, \ldots,$ and Wip_i for an integer p_i;
  else $W_{i1} = W_i$;
  end if
end for
sort data warehouses on size in non-increasing order and the data warehouses are renamed as $DW_1, DW_2, \ldots, DW_N$, where $N = \sum_{i=1}^{n} p_i$;
let $LPB_0 = \phi$;
for $i = 1$ to N **do**
  mine DW_i using a SDMT with input LPB_{i-1};
end for
return $LPB_1, LPB_2, \ldots, LPB_N$;
end procedure

In the algorithm, the usage of LPB_{i-1} during mining DW_i has been explained above. Once a pattern has been extracted from a data warehouse, then it also gets extracted from the remaining data warehouses. Thus, the algorithm *PipelinedFeedbackModel* improves the quality of synthesized patterns as well as contributes significantly to an analysis of local patterns.

3.6 Error Evaluation

To evaluate the quality of MDMT: PFM+SPS, one needs to quantify the error produced by the method. First, in an experiment we mine frequent itemsets in multiple databases using PFM, and afterwards synthesize global patterns using the SPS algorithm. One needs to find how the global synthesized support differs from the exact (apriori) support of an itemset.

PFM improves mining multiple databases significantly over local pattern analysis. In the PFM, we have $LPB_{i-1} \subseteq LPB_i$, for $i = 2, 3, \ldots, n$. Then, patterns in $LPB_i - LPB_{i-1}$ are generated from databases $D_i, D_{i+1}, \ldots, D_n$. We assume $supp_a(X, D_j) = 0$, for each $X \in LPB_i - LPB_{i-1}$, and $j = 1, 2, \ldots, i-1$. Thus, the error of mining X could be defined as follows:

$$\mathrm{E}(X|\mathrm{PFM, SPS}) = \left| supp_a(X, D) - \frac{1}{\sum_{j=1}^{n} |D_j|} \times \sum_{j=i}^{n} \left[supp_a(X, D_j) \times |D_j| \right] \right|, \tag{3.6}$$

$$\text{for } X \in LPB_i - LPB_{i-1} \text{ and } i = 2, 3, \ldots, n.$$

Also, $\mathrm{E}(X|\mathrm{PFM, SPS}) = 0$, for $X \in LPB_1$.

When a frequent itemset is reported from D_1 then it gets reported from every databases using PFM algorithm. Thus, $\mathrm{E}(X|\mathrm{PFM, SPS}) = 0$, for $X \in LPB_1$.

Otherwise, an itemset X is not reported from all the databases. It is synthesized using SPS algorithm. Then the synthesized support is subtracted from its apriori support for finding the error of mining X.

There are several ways one could define the error of an experiment. In particular, one could concentrate on the following definitions.

1. *Average error* (AE)

$$\mathrm{AE}(\mathrm{D}, \alpha) = \frac{1}{|LPB_1 + \sum_{i=2}^{n}(LPB_i - LPB_{i-1})|} \sum_{X \in [LPB_1 \cup \{\cup_{i=2}^{n}(LPB_i - LPB_{i-1})\}]} \mathrm{E}(X|\mathrm{PFM, SPS})] \tag{3.7}$$

2. *Maximum error* (ME)

$$\mathrm{ME}(\mathrm{D}, \alpha) = maximum \left\{ \mathrm{E}(X|\mathrm{PFM, SPS}), \text{for} X \in LPB_1 \cup \left\{ \cup_{i=2}^{n}(LPB_i - LBP_{i-1}) \right\} \right\} \tag{3.8}$$

where $supp_a(X_i, D)$ is obtained by mining D using a traditional data mining technique, $i = 1, 2, \ldots, m$. $supp_s(X_i, \mathrm{D})$ is obtained by SPS, for $i = 1, 2, \ldots, m$.

3.7 Experiments

We have carried out a series of experiments to study and quantify the effectiveness of the PFM. We present experimental results using three synthetic databases and two real-world databases. The synthetic databases are *T10I4D100K* (*T*) (Frequent itemset mining dataset repository 2004), *random500* (*R*1) and *random1000* (*R*2). The databases *random500* and *random1000* are generated synthetically for the purpose of conducting experiments. The real databases are *retail* (*R*) (Frequent itemset mining dataset repository 2004) and *BMS-Web-Wiew-*1 (*B*) (Frequent itemset mining dataset repository 2004). The main characteristics of these datasets are displayed in Table 3.1.

Table 3.1 Database characteristics

D	*NT*	*ALT*	*AFI*	*NI*
T	1,00,000	11.10	1,276.12	870
R	88,162	11.31	99.67	10,000
B	1,49,639	2.00	155.71	1,922
*R*1	10,000	6.47	109.40	500
*R*2	10,000	12.49	111.86	1,000

Let *NT*, *AFI*, *ALT*, and *NI* denote the number of transactions, average frequency of an item, average length of a transaction, and number of items in a database, respectively. Each of the above databases is split into 10 databases for the purpose of carrying out experiments. The databases obtained from *T*, *R*, *B*, *R*1 and *R*2 are named as T_i, R_i, B_i, $R1_i$ and $R2_i$, respectively, for $i = 0, 1, \ldots, 9$. The databases T_i, R_i, B_i, $R1_i$, $R2_i$ are called input databases (*DB*s), for $i = 0, 1, \ldots, 9$. Some characteristics of these input databases are presented in the Table 3.2. In Tables 3.3 and 3.4, we include some outcomes to quantify how the proposed technique improves the results of mining. We have completed experiments using other MDMTs on these databases for the purpose of comparing them with MDMT: PFM+SPS.

Figures 3.3, 3.4, 3.5, 3.6 and 3.7 show average error versus different values of α. From these graphs, we conclude that AE normally increases as α increases. The number of databases reporting a pattern decreases when the values of α increase. Thus, the AE of synthesizing patterns normally increases as α increases. In case of Fig. 3.5, the graphs for MDMT: PFM+SPS and MDMT: RO+PA are similar to those with the X-axis.

Table 3.2 Input database characteristics

D	*NT*	*ALT*	*AFI*	*NI*	*DB*	*NT*	*ALT*	*AFI*	*NI*
T_0	10,000	11.06	127.66	866	T_5	10,000	11.14	128.63	866
T_1	10,000	11.133	128.41	867	T_6	10,000	11.11	128.56	864
T_2	10,000	11.07	127.64	867	T_7	10,000	11.10	128.45	864
T_3	10,000	11.12	128.44	866	T_8	10,000	11.08	128.56	862
T_4	10,000	11.14	128.75	865	T_9	10,000	11.08	128.11	865
R_0	9,000	11.24	12.07	8,384	R_5	9,000	10.86	16.71	5,847
R_1	9,000	11.21	12.27	8,225	R_6	9,000	11.20	17.42	5,788
R_2	9,000	11.34	14.60	6,990	R_7	9,000	11.16	17.35	5,788
R_3	9,000	11.49	16.66	6,206	R_8	9,000	12.00	18.69	5,777
R_4	9,000	10.96	16.04	6,148	R_9	7,162	11.69	15.35	5,456
B_0	14,000	2.00	14.94	1,874	B_5	14,000	2.00	280.00	100
B_1	14,000	2.00	280.00	100	B_6	14,000	2.00	280.00	100
B_2	14,000	2.00	280.00	100	B_7	14,000	2.00	280.00	100
B_3	14,000	2.00	280.00	100	B_8	14,000	2.00	280.00	100
B_4	14,000	2.00	280.00	100	B_9	23,639	2.00	472.78	100
$R1_0$	1,000	6.37	10.73	500	$R1_5$	1,000	6.34	10.68	500
$R1_1$	1,000	6.50	11.00	500	$R1_6$	1,000	6.62	11.25	500
$R1_2$	1,000	6.40	10.80	500	$R1_7$	1,000	6.42	10.83	500
$R1_3$	1,000	6.52	11.05	500	$R1_8$	1,000	6.58	11.16	500
$R1_4$	1,000	6.30	10.60	500	$R1_9$	1,000	6.65	11.30	500
$R2_0$	1,000	6.42	5.43	996	$R2_5$	1,000	6.44	5.46	997
$R2_1$	1,000	6.41	5.44	995	$R2_6$	1,000	6.48	5.50	996
$R2_2$	1,000	6.56	5.58	995	$R2_7$	1,000	6.48	5.49	997
$R2_3$	1,000	6.53	5.54	998	$R2_8$	1,000	6.54	5.56	996
$R2_4$	1,000	6.50	5.54	991	$R2_9$	1,000	6.50	5.56	988

Table 3.3 Error obtained for the first three databases for selected value of α

Database	*T10I4D100K*		*retail*		*BMS-Web-Wiew-1*	
α	0.05		0.11		0.19	
Error type	AE	ME	AE	ME	AE	ME
MDMT: RO+IEP	0.01	0.04	0.01	0.06	0.05	0.15
MDMT: RO+RS	0.01	0.04	0.01	0.06	0.02	0.13
MDMT: RO+ARS	0.01	0.04	0.01	0.06	0.02	0.11
MDMT: PFM+SPS	0	0.05	0.01	0.06	0	0
MDMT: RO+PA	0	0	0	0	0	0

3.8 Conclusions

In this chapter, we have discussed several generalized as well as specialized multi-database mining techniques. For a particular problem at hand, one technique could be more suitable than the others. However, we cannot claim that there is a single

Table 3.4. Error reported for the last two databases for selected value of α

Database	*random500*		*random1000*	
α	0.005		0.004	
Error type	AE	ME	AE	ME
MDMT: RO+IEP	0.01	0.01	0.01	0.01
MDMT: RO+RS	0.01	0.01	0	0.01
MDMT: RO+ARS	0.01	0.01	0.01	0.01
MDMT: PFM+SPS	0.01	0.01	0	0
MDMT: RO+PA	0	0	0	0

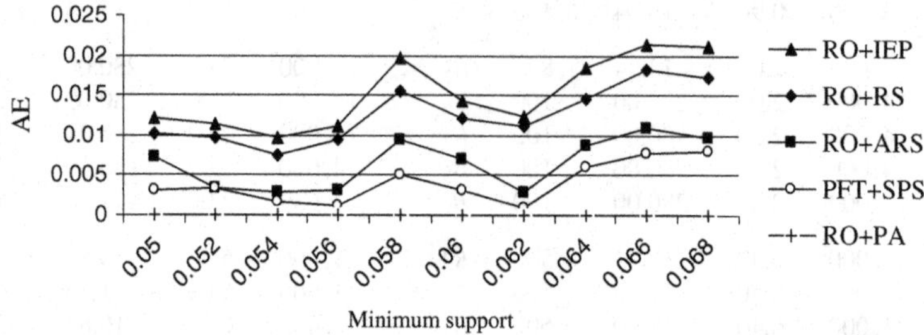

Fig. 3.3 AE vs. α for experiments conducted for database T

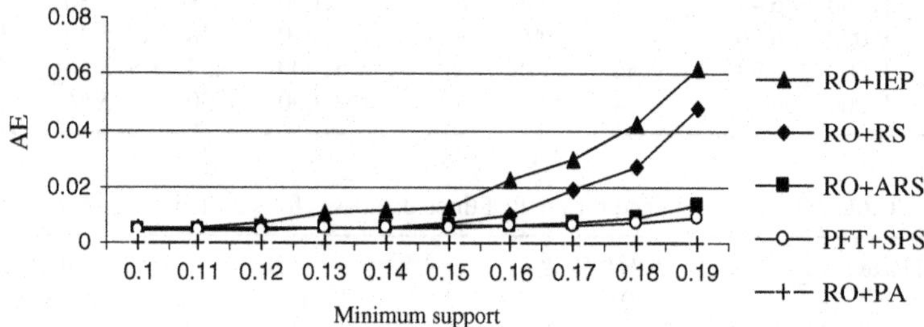

Fig. 3.4 AE vs. α for experiments using database R

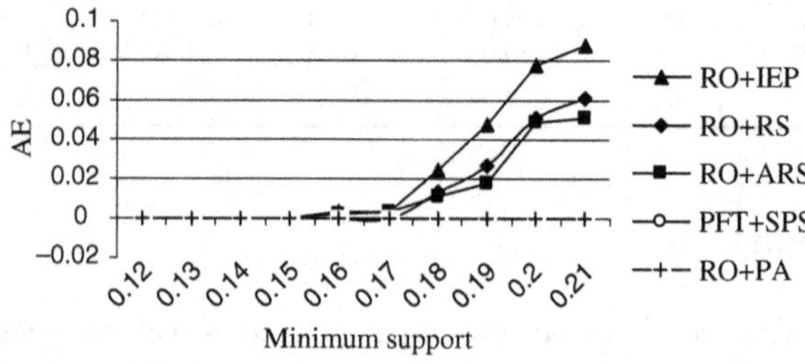

Fig. 3.5 AE vs. α for experiments using database B

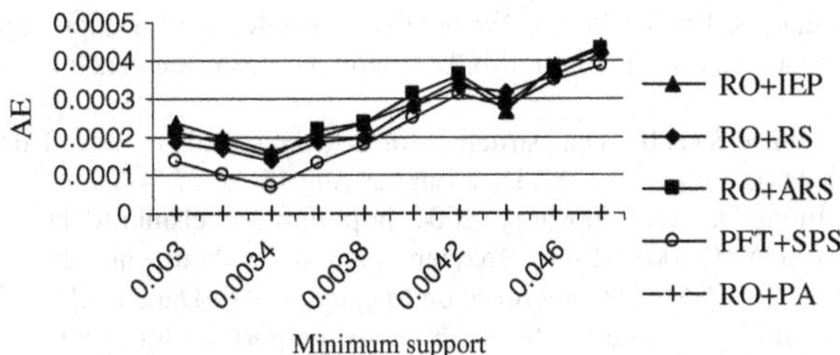

Fig. 3.6 AE vs. α for experiments using database $R1$

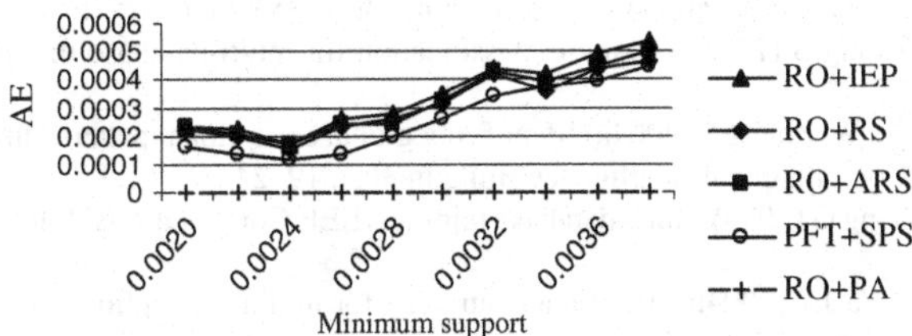

Fig. 3.7 AE vs. α for experiments using database $R2$

method of universal nature which outperforms all other techniques. Instead, a choice of the method has to be problem-driven. We have formalized the idea of multi-database mining using local pattern analysis by considering an underlying two-step process. We have also presented the pipelined feedback model which is particularly suitable for mining multiple large databases. It improves significantly the accuracy of mining multiple databases as compared to an existing technique that scans each database only once. The pipelined feedback model could also be used for mining a large database by dividing it into a series of sub-databases. Experimental results obtained with the use of the MDMT: PFM+SPS are promising and underline the usefulness of the method studied here.

References

Adhikari A, Rao PR (2007a) Study of select items in multiple databases by grouping. In: Proceedings of the 3rd Indian International Conference on Artificial Intelligence, pp. 1699–1718

Adhikari A, Rao PR (2008a) Synthesizing heavy association rules from different real data sources. Pattern Recognition Letters 29(1): 59–71

Adhikari A, Rao PR (2008b) Efficient clustering of databases induced by local patterns. Decision Support Systems 44(4): 925–943

Adhikari J, Rao PR (2009) Clustering items in different data sources induced by stability. The International Arab Journal of Information Technology 6(4): 66–74

Adhikari A, Rao PR, Adhikari J (2007b) Mining multiple large databases. In: Proceedings of the 10th International Conference on Information Technology, Washington, DC, pp. 80–84

Agrawal R, Srikant R (1994) Fast algorithms for mining association rules. In: Proceedings of International Conference on Very Large Data Bases, pp. 487–499

Babcock B, Chaudhury S, Das G (2003) Dynamic sample selection for approximate query processing. In: Proceedings of ACM SIGMOD Conference Management of Data, New York, pp. 539–550

Coenen F, Leng P, Ahmed S (2004) Data structure for association rule mining: T-trees and P-trees. IEEE Transactions on Knowledge and Data Engineering 16(6): 774–778

Frequent Itemset Mining Dataset Repository (2004) http://fimi.cs.helsinki.fi/data

Han J, Pei J, Yiwen Y (2000) Mining frequent patterns without candidate generation. In: Proceedings of ACM SIGMOD Conference on Management of Data, Dallas, TX, pp. 1–12

Savasere A, Omiecinski E, Navathe S (1995) An efficient algorithm for mining association rules in large databases. In: Proceedings of the 21st International Conference on Very Large Data Bases, Zurich, Switzerland, pp. 432–443

Wu X, Zhang S (2003) Synthesizing high-frequency rules from different data sources. IEEE Transactions on Knowledge and Data Engineering 14(2): 353–367

Wu X, Zhang C, Zhang S (2005) Database classification for multi-database mining. Information Systems 30(1): 71–88

Zhang C, Liu M, Nie W, Zhang S (2004a) Identifying global exceptional patterns in multi-database mining. IEEE Computational Intelligence Bulletin 3(1): 19–24

Zhang S, Wu X, Zhang C (2003) Multi-database mining. IEEE Computational Intelligence Bulletin 2(1): 5–13

Zhang S, Zhang C, Yu JX (2004b) An efficient strategy for mining exceptions in multi-databases. Information Sciences 165(1–2): 1–20

Chapter 4
Mining Patterns of Select Items in Multiple Databases

A number of important decisions are based on a set of specific items in a database called the *select items*. Thus the analysis of select items in multiple databases becomes of primordial relevance. In this chapter, we focus on the following issues. First, a model of mining global patterns of select items from multiple databases is presented. Second, a measure of quantifying an overall association between two items in a database is discussed. Third, we present an algorithm that is based on the proposed overall association between two items in a database for the purpose of grouping the frequent items in multiple databases. Each group contains a select item called the *nucleus item* and the group grows while being centered around the nucleus item. Experimental results are concerned with some synthetic and real-world databases.

4.1 Introduction

In Chapter 3, we have presented a generalized technique viz., MDMT: PFM+SPS, for mining multiple large databases. We have noted that one could develop a multi-database mining application using MDMT: PFM+SPS which performs reasonably well. The following question arises as to whether MDMT: PFM+SPS is the most suitable technique for mining multiple large databases in all situations. In many applications, one may need to extract true non-local patterns of a set of specific items present in multiple large databases. In such applications, MDMT: PFM+SPS could not be suggested as it may return approximate non-local patterns. In this chapter, we present a technique that extracts genuine global patterns of a set of specific items from multiple large databases.

Many decisions are based on a set of specific items called *select items*. Let us highlight several decision support applications where the decisions are based on the performance of select items.

- Consider a set of items (products) that are profit making. We could consider them as the select items in this context. Naturally, the company would like to promote them. There are various ways one could promote an item. An indirect way of promoting a select item is to promote items that are positively associated with

A. Adhikari et al., *Developing Multi-database Mining Applications*, Advanced Information and Knowledge Processing, DOI 10.1007/978-1-84996-044-1_4,

it. The implication of positive association between a select item P and another item Q is that if Q is purchased by a customer then P is likely to be purchased by the same customer at the same time. In this way, item P becomes indirectly promoted. It is important to identify the items that are positively associated with a select item.
- Each of the select items could be of high standard. Thus, they bring goodwill for the company. They help promoting other items. Therefore it is essential to know how the sales of select items affect the other items. Before proceeding with such analyses, one may need to identify the items that are positively associated with the select items.
- Again, each of the select items could be a low-profit making product. From this perspective, it is important to know how they promote the sales of other items. Otherwise, the company could stop dealing with those products.

In general, the performance of select items could affect many decision making problems. Thus a better, more comprehensive analysis of select items might lead to better decisions. We study the select items based on the frequent itemsets extracted from multiple databases. The first question is whether a "traditional" data mining technique could provide a good solution when dealing with multiple large databases. The "traditional" way of mining multiple databases might not provide a good solution due to several reasons:

- The company might have to employ parallel hardware and software to deal with a large volume of data.
- A single computer might take unreasonable amount of time to mine a large volume of data. In some extreme cases, it might not be feasible to carry data mining.
- A traditional data mining algorithm might extract a large number of patterns comprising many irrelevant items. Thus the processing of patterns could be complex and time consuming.

Therefore, the traditional way of mining multiple databases could not provide an efficient solution to the problem. In this situation, one could apply local pattern analysis (Zhang et al. 2003). Given this model of mining multiple databases, each branch of a company requires to mine the local database by utilizing some traditional data mining technique. Afterwards, each branch forwards the pattern base to the central office. The central office processes such pattern bases collected from different branches and synthesizes the global patterns and eventually makes decisions. Due to the reasons stated above, the local pattern analysis would not be a judicious choice to solve the proposed problem.

Each local pattern base might contain a large number of patterns consisting of many irrelevant items. Under these circumstances, the data analysis becomes complicated and time consuming. A pattern of a select item might be absent in some local pattern bases. One may be required to estimate or ignore some patterns in certain databases. Therefore we may fail to report the true global patterns of select items in the union of all local databases. All in all, we conclude that the local pattern analysis alone might not provide a good solution to the problem.

Due to difficulties identified above, we aim at developing a technique that mines *true* global patterns of select items in multiple databases. There are two apparent advantages of using such technique. First, the synthesized global patterns are exact. In other words, there is no necessity to estimate some patterns in some databases. Second, we avoid dealing with huge volumes of data.

4.2 Mining Global Patterns of Select Items

In Fig. 4.1, we show an essence of the technique of mining global patterns of select items in multiple databases (Adhikari and Rao 2007). It consists of the following steps:

1. Each branch constructs the database and sends it to the central office.
2. Also, each branch extracts patterns from its local database.
3. The central office amalgamates these forwarded databases into a single database *FD*.
4. A traditional data mining technique is applied to extract patterns from *FD*.
5. The global patterns of select items could be extracted effectively from local patterns and the patterns extracted from *FD*.

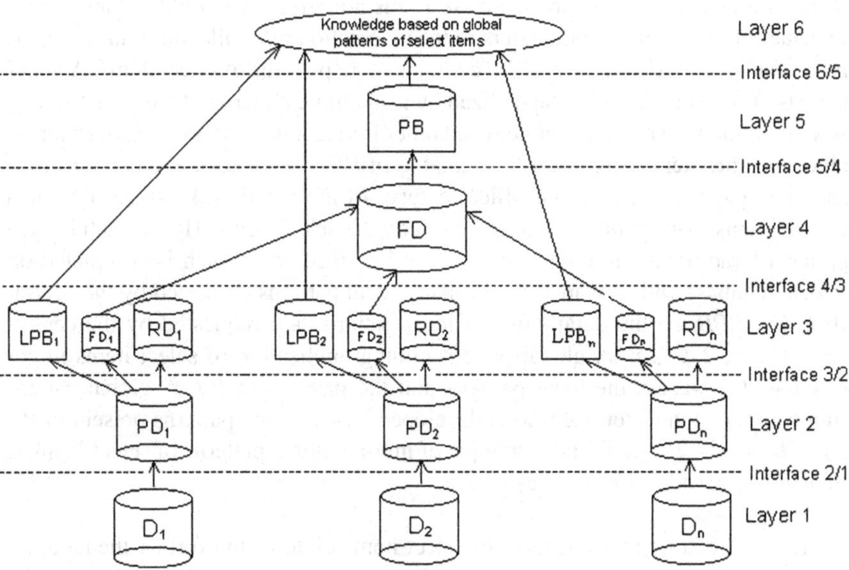

Fig. 4.1 A multilevel process of mining global patterns of select items in multiple databases

In Section 4.4, we will explain steps 1–5 with the help of a specific illustrative example. The local databases are located at the bottom level of the figure. We need to process these databases as they may not be at the appropriate state for the mining task. Various data preparation techniques (Pyle 1999) like data cleaning, data transformation, data integration, data reduction etc. are applied to these data present in local databases. We produce local processed database PD_i for the i-th branch, for $i = 1, 2, \ldots, n$. The proposed model comes with a set of interfaces combined with a set of layers. Each interface forms a set of operations that produces dataset(s) (or knowledge) based on the dataset(s) available at the lower level. There are five interfaces in the proposed model. The functions of the interfaces are described below.

Interface 2/1 is used to clean/transform/integrate/reduce data present at the lowest level. By applying these procedures we construct database resulting from the original database. These operations are carried out at the respective branch. We apply an algorithm (located at interface 3/2) to partition a local database into two parts: forwarded database and remaining database. It is easy to find the forwarded database corresponding to a given database. In the following paragraph, we discuss how to construct FD_i, from D_i, for $i = 1, 2, \ldots, n$.

Initially, FD_i is kept empty. Let T_{ij} be the j-the transaction of D_i, $j = 1, 2, \ldots, |D_i|$. For D_i, a for-loop on j would run $|D_i|$ times. At the j-th iteration, the transaction T_{ij} is tested. If T_{ij} contains at least one of the select items then FD_i is updated, resulting in the union $FD_i \cup \{T_{ij}\}$. At the end of the for-loop completed for j, FD_i is constructed.

A transaction related to select items might contain items other than those being selected. A traditional data mining algorithm could be applied to extract patterns from *FD*. Let *PB* be the pattern base returned by a traditional data mining algorithm (at the interface 5/4). Since the database *FD* is not large, one could reduce further the values of user-defined parameters of the association rules, like minimum support and minimum confidence, so that *PB* contains more patterns of select items. A better analysis of select items could be realized by using more patterns. If we wish to study the association between a select item and other frequent items then the exact support values of other items might not be available in *PB*. In this case, the central office sends a request to each branch office to forward the details (like support values) of some items that would be required to study the select items. Hence each branch applies a "traditional" mining algorithm (at interface 3/2) which is completed on its local database and forwards the details of local patterns requested by the central office. Let LPB_i be the details of i-th local pattern base requested by the central office, $i = 1, 2, \ldots, n$. A global pattern mining application of select items might be required to access the local patterns and the patterns in *PB*. A global pattern mining application (interface 6/5) is developed based on the patterns present in *PB* and LPB_i, $i = 1, 2, \ldots, n$. The technique of mining global patterns of select items is efficient due to the following reasons:

- One could extract more patterns of select items by lowering further the parameters of association rule such as the minimum support and minimum confidence, based on the level of data analysis of select items, since *FD* is reasonably small.
- We get true global patterns of select items as there is no need to estimate them.

In light of these observations, we can anticipate that the quality of global patterns is high, since there is no need to estimate them.

To evaluate the effectiveness of the above technique, we present a problem on multi-database mining. We show how the data mining technique presented above could be used in finding the solution to the problem. We start with the notion of overall association between two items in a database (Adhikari and Rao 2007).

4.3 Overall Association Between Two Items in a Database

Let *I(DB)* be the set of items in database *DB*. A common measure of similarity (Wu et al. 2005; Xin et al. 2005) between two objects could be used as a measure of positive association between two items in a database. We define positive association between two items in a database as follows:

$$PA(x, y, DB) = \frac{\#\,\text{transaction containing both } x \text{ and } y,\ DB}{\#\,\text{transaction containing at least one of } x \text{ and } y,\ DB},\ \text{for } x, y \in I(DB) \tag{4.1}$$

where, # *P*, *DB* is the number of transactions in *DB* that satisfy predicate *P*. *PA* measures only positive association between two items in a database. It does not measure negative association between two items in a database. In the following example, we show that *PA* fails to compute an overall association between two items.

Example 4.1 Let us consider four branches of a multi-branch company. Let D_i be the database corresponding to the *i*-th branch of the company, $i = 1, 2, 3, 4$. The company is interested in analyzing globally a set of select items (*SI*). Let $SI = \{a, b\}$. The contents of different databases are given as follows: $D_1 = \{ \{a, e\}, \{b, c, g\}, \{b, e, f\}, \{g, i\} \}$; $D_2 = \{ \{b, c\}, \{f, h\} \}$; $D_3 = \{ \{a, b, c\}, \{a, e\}, \{c, d\}, \{g\} \}$; $D_4 = \{ \{a, e\}, \{b, c, g\} \}$. Initially, we wish to measure the association between two items in a single database, say D_1. Now, $PA(a, b, D_1) = 0$, since there is no transaction in D_1 containing both the items *a* and *b*. In these transactions, if one of the items of $\{a, b\}$ is present then the other item of $\{a, b\}$ is not present. Thus, the transactions $\{a, e\}$, $\{b, c, g\}$ and $\{b, e, f\}$ in D_1 imply that the items *a* and *b* are negatively associated. We need to define a measure of negative association between two items in a database. Similarly to the measure of positive association, one could define a measure of negative association between two items in a database as follows:

$$NA(x, y, DB) = \frac{\#\,\text{transaction containing exactly one of } x \text{ and } y,\ DB}{\#\,\text{transaction containing at least one of } x \text{ and } y,\ DB}\ \text{for } x,\ y \in I(DB). \tag{4.2}$$

Now, $NA(a, b, D_1) = 1$. We note that $PA(a, b, D_1) < NA(a, b, D_1)$. Overall, we state that the items *a* and *b* are negatively associated, and the amount of overall association between the items *a* and *b* in D_1 is $PA(a, b, D_1) - NA(a, b, D_1) = -1.0$. The accuracy of association analysis might be low if we consider only the positive association between two items.

The analysis of relationships among variables is a fundamental task being at the heart of many data mining problems. For example, metrics such as support, confidence, lift, correlation, and collective strength have been used extensively to evaluate the interestingness of association patterns (Klemettinen et al. 1994; Silberschatz and Tuzhilin 1996; Aggarwal and Yu 1998; Silverstein et al. 1998; Liu et a. 1999). These metrics are defined in terms of the frequency counts tabulated in a 2 × 2 contingency table as shown in Table 4.1. Tan et al. (2002) presented an overview of twenty one interestingness measures proposed in statistics, machine learning and data mining literature. We continue our discussion with the examples cited in Tan et al. (2002) and show that none of the proposed measures is effective in finding the overall association by considering both positive and negative associations between two items in a database.

Table 4.1 A 2 × 2 contingency table for variables x and y

	y	$\neg y$	Total
x	f_{11}	f_{10}	$f_{1.}$
$\neg x$	f_{01}	f_{00}	$f_{0.}$
Total	$f_{.1}$	$f_{.0}$	$f_{..}$

Table 4.2 Examples of contingency tables

Example	f_{11}	f_{10}	f_{01}	f_{00}
E1	8,123	83	424	1,370
E2	8,330	2	622	1,046
E3	9,481	94	127	298
E4	3,954	3,080	5	2,961
E5	2,886	1,363	1,320	4,431
E6	1,500	2,000	500	6,000
E7	4,000	2,000	1,000	3,000
E8	4,000	2,000	2,000	2,000
E9	1,720	7,121	5	1,154
E10	61	2,483	4	7,452

From the examples in Table 4.2, we notice that the overall association between two items could be negative as well as positive. In fact, a measure of overall association between two items in a database lies in [−1, 1]. We consider the following five out of 21 interestingness measures, since the association between two items calculated using one of these five measures lies in [−1, 1]. Thus, we study their usefulness for the specific requirement of the proposed problem. These five measures are included in Table 4.3.

Table 4.3 Selected interestingness measures for association patterns

Symbol	Measure	Formula
ϕ	ϕ-coefficient	$\frac{P(\{x\}\cup\{y\})-P(\{x\})\times P(\{y\})}{\sqrt{P(\{x\})\times P(\{y\})\times(1-P(\{x\})\times(1-P(\{y\}))}}$
Q	Yule's Q	$\frac{P(\{x\}\cup\{y\})\times P(\neg(\{x\}\cap\{y\}))-P(\{x\}\cup\neg\{y\})\times P(\neg\{x\}\cup\{y\})}{P(\{x\}\cup\{y\})\times P(\neg(\{x\}\cap\{y\}))-P(\{x\}\cup\neg\{y\})\times P(\neg\{x\}\cup\{y\})}$
Y	Yule's Y	$\frac{\sqrt{P(\{x\}\cup\{y\})\times P(\neg(\{x\}\cap\{y\}))}-\sqrt{P(\{x\}\cup\neg\{y\})\times P(\neg\{x\}\cup\{y\})}}{\sqrt{P(\{x\}\cup\{y\})\times P(\neg(\{x\}\cap\{y\}))}-\sqrt{P(\{x\}\cup\neg\{y\})\times P(\neg\{x\}\cup\{y\})}}$
κ	Cohen's κ	$\frac{P(\{x\}\cup\{y\})+P(\neg\{x\}\cup\neg\{y\})-P(\{x\})\times P(\{y\})-P(\neg\{x\})\times P(\neg\{y\})}{1-P(\{x\})\times P(\{y\})-P(\neg\{x\})\times P(\neg\{y\})}$
F	Certainty factor	$\max\left(\frac{P(\{y\}\vert\{x\})-P(\{y\})}{1-P(\{y\})}, \frac{P(\{x\}\vert\{y\})-P(\{x\})}{1-P(\{x\})}\right)$

In Table 4.4, we rank the contingency tables by using each of the above measures.

Table 4.4 Ranking of contingency tables using above interestingness measures

Example	*E1*	*E2*	*E3*	*E4*	*E5*	*E6*	*E7*	*E8*	*E9*	*E10*
ϕ	1	2	3	4	5	6	7	8	9	10
Q	3	1	4	2	8	7	9	10	5	6
Y	3	1	4	2	8	7	9	10	5	6
κ	1	2	3	5	4	7	6	8	9	10
F	4	1	6	2	9	7	8	10	3	5

Also, we rank the contingency tables based on the concept of overall association explained in Example 4.1. In Table 4.5, we present the ranking of contingency tables using overall association.

Table 4.5 Ranking of contingency tables using overall association

Example	Overall association	Rank
E1	0.76	3
E2	0.77	2
E3	0.93	1
E4	0.09	5
E5	0.02	6
E6	−0.10	8
E7	0.10	4
E8	0	7
E9	−0.54	10
E10	−0.24	9

The ranks given in Table 4.5 and the ranks given for each of the five measures in Table 4.4 are not similar. In other words, none of the above five measures satisfies the requirement formulated in the proposed problem. Based on the above discussion, we propose the following measure *OA* as an overall association between two items in a database.

Definition 4.1 $OA(x, y, DB) = PA(x, y, DB) - NA(x, y, DB)$, for $x, y \in I(DB)$.

If $OA(x, y, DB) > 0$ then the items x and y are positively associated in DB. If $OA(x, y, DB) < 0$ then the items x and y are negatively associated in DB. The problem is concerned with the association between a nucleus item and another item in a database. Thus, we are not concerned about the association between two items in a group, where none of them is a nucleus item. In other words, it could be considered as a problem of grouping rather than a problem of classification or clustering.

4.4 An Application: Study of Select Items in Multiple Databases Through Grouping

As before, let us consider a multi-branch company having n branches. Each branch maintains a separate database for the transactions made in that particular branch. Let D_i be the database corresponding to the i-th branch of the multi-branch company, $i = 1, 2, \ldots, n$. Also, let D be the union of all branch databases. A large section of a local database might be irrelevant to the current problem. Thus, we divide database D_i into FD_i and RD_i, where FD_i and RD_i are called the *forwarded database* and *remaining database* corresponding to the i-th branch, respectively, $i = 1, 2, \ldots, n$. We are interested in the forwarded databases, since every transaction in a forwarded database contains at least one select item. The database FD_i is forwarded to the central office for mining global patterns of select items, for $i = 1, 2, \ldots, n$. All the local forwarded databases are amassed into a single database (FD) for the purpose of mining task. We note that the database FD is not overly large as it contains transactions related to select items. Before proceeding with the detailed discussion, we first offer some definitions.

A set of items is referred to as an *itemset*. An itemset containing k items is called a *k-itemset*. The *support* (*supp*) (Agrawal et al. 1993) of an itemset refers to the fraction of transactions containing this itemset. If an itemset satisfies the user-specified minimum support (α) criterion, then it is called a *frequent itemset* (*FIS*). Similarly, if an item satisfies the user-specified minimum support criterion, then it is called a *frequent item* (*FI*). If a k-itemset is frequent then every item in the k-itemset is also frequent. In this chapter, we study the items in SI. Let $SI = \{s_1, s_2, \ldots, s_m\}$. We wish to construct m groups of frequent items in such a way that the i-th group grows by being centered around the nucleus item s_i, $i = 1, 2, \ldots, m$. Let FD be the union of FD_i, $i = 1, 2, \ldots, n$. Furthermore let $FIS_k(DB \mid \alpha)$ be the set of frequent k-itemsets in database DB at the minimum support level α, $k = 1, 2$. We state our problem as follows:

Let G_i be the i-the group of frequent items containing the nucleus item $s_i \in SI$, $i = 1, 2, \ldots, m$. Construct G_i using $FIS_2(FD \mid \alpha)$ and local patterns in D_i such that $x \in G_i$ implies $OA(s_i, x, D) > 0$, for $i = 1, 2, \ldots, m$.

Two groups may not be mutually exclusive, as our study involves identifying pairs of items such that the following conditions are true: (i) the items in each pair are positively associated between each other in D, and (ii) one of the items in a pair is a select item. Our study is not concerned with the association between a pair of items in a group such that none of them is a select item. The above problem actually results in m +1 groups where (m +1)-th group G_{m+1} contains the items that are not positively associated with any one of the select items. The proposed study is not concerned with the items in G_{m+1}.

The crux of the proposed problem is to determine the supports of the relevant frequent itemsets in multiple large databases. A technique of estimating support of a frequent itemset in multiple real databases has been proposed by Adhikari and Rao (2008a). To estimate the support of an itemset in a database, this technique makes use of the trend of supports of the same itemset in other databases. The trend approach for estimating support of an itemset in a database could be described as follows:

Let the itemset X gets reported from databases $D_1, D_2, \ldots, D_m$. Also let *supp*(X, $\cup_{i=1}^{m} D_i$) be the support of X in the union of $D_1, D_2, \ldots, D_m$. Let D_k be a database that does not report X, for $k = m + 1, m + 2, \ldots, n$. Then the support of X in D_k could be estimated by $\alpha \times$ *supp*(X, $\cup_{i=1}^{m} D_i$). Given an itemset X, some local supports of X are estimated and the remaining local supports of X are obtained using a traditional data mining technique. The global support of X is obtained by combining these local supports with the numbers of transactions (i.e., sizes) of the respective databases. The proposed technique synthesizes true supports of relevant frequent itemsets in multiple databases.

We have discussed the limitations of the traditional way of mining multiple large databases in the previous chapters. We have observed that local pattern analysis alone could not provide an effective solution to this problem. The mining technique visualized in Fig. 4.1 offers a viable solution. A pattern based on all the databases is called a *global pattern*. A global pattern containing at least one select item is called a *global pattern of select item*.

4.4.1 Properties of Different Measures

If the itemset $\{x, y\}$ is frequent in *DB* then *OA*(x, y, *DB*) is not necessarily be positive, since the number of transactions containing only one of the items of $\{x, y\}$ could be more than the number of transactions containing both the items x and y. *OA*(x, y, *DB*) could attain maximum value for an infrequent itemset $\{x, y\}$ also. Let $\{x, y\}$ be infrequent. The distributions of x and y in *DB* are such that no transaction in *DB* contains only one item of $\{x, y\}$. Thus, *OA*(x, y, *DB*) $= 1.0$. In what follows, we discuss a few properties of different measures.

Lemma 4.1 (i) $0 \le PA(x, y, DB) \le 1$; (ii) $0 \le NA(x, y, DB) \le 1$; (iii) $-1 \le OA(x, y, DB) \le 1$; (iv) $PA(x, y, DB) + NA(x, y, DB) = 1$; *for* $x, y \in I(DB)$.

PA(x, y, *DB*) could be considered as a similarity between x and y in *DB*. Thus, 1− *PA*(x, y, *DB*) i.e., *NA*(x, y, *DB*) could be considered as a distance between x and y in *DB*. A characteristic of a good distance measure is that it satisfies metric properties (Barte 1976) over the concerned domain.

Lemma 4.2 *$NA(x, y, DB) = 1- PA(x, y, DB)$ is a metric over* $[0, 1]$, *for* $x, y \in I(DB)$.

Proof We prove only the property of triangular inequality, since the remaining two properties of the metric are obvious. Let $I(DB) = \{a_1, a_2, \ldots, a_N\}$. Let ST_i be the set of transactions containing item $a_i \in I(DB)$, $i = 1, 2, \ldots, N$.

$$1-PA(a_p, a_q, DB) = 1-\frac{|ST_p \cap ST_q|}{|ST_p \cup ST_q|} \geq \frac{|ST_p-ST_q| + |ST_q-ST_p|}{|ST_p \cup ST_q \cup ST_r|} \tag{4.3}$$

Thus, $1- PA(a_p, a_q, DB) + 1- PA(a_q, a_r, DB)$

$$\geq \frac{|ST_p-ST_q| + |ST_q-ST_p| + |ST_q-ST_r| + |ST_r-ST_q|}{|ST_p \cup ST_q \cup ST_r|} \tag{4.4}$$

$$= \frac{|ST_p \cup ST_q \cup ST_r| - |ST_p \cap ST_q \cap ST_r| + |ST_p \cap ST_r| + |ST_q| - |ST_p \cap ST_q| - |ST_q \cap ST_r|}{|ST_p \cup ST_q \cup ST_r|} \tag{4.5}$$

$$= 1-\frac{|ST_p \cap ST_q \cap ST_s| - |ST_p \cap ST_s| - |ST_q| + |ST_p \cap ST_q| + |ST_q \cap ST_s|}{|ST_p \cup ST_q \cup ST_s|} \tag{4.6}$$

$$= 1-\frac{\{|ST_p \cap ST_q \cap ST_s| + |ST_p \cap ST_q| + |ST_q \cap ST_s|\} - \{|ST_p \cap ST_s| + |ST_q|\}}{|ST_p \cup ST_q \cup ST_s|} \tag{4.7}$$

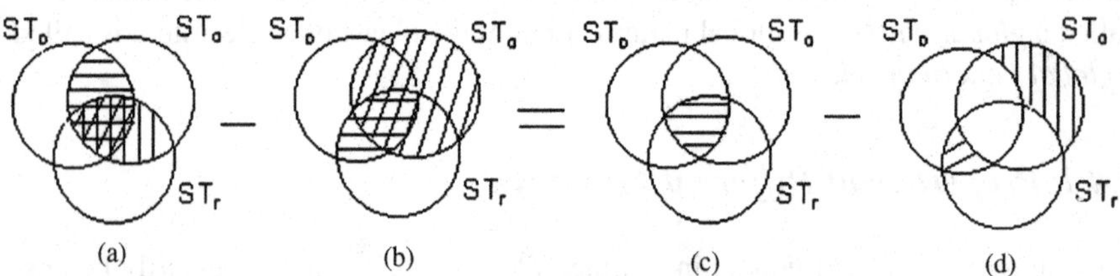

Fig. 4.2 Simplification using Venn diagram

Let the number of elements in the shaded region of Figs. 4.2(c) and 4.2(d) be N_1 and N_2, respectively. Then the expression (4.7) becomes

$$1-\frac{N_1-N_2}{|ST_p \cup ST_q \cup ST_r|} \geq \begin{cases} 1-\dfrac{N_1-N_2}{|ST_p \cup ST_q \cup ST_r|}, \text{if } N_1 \geq N_2 \text{ (case 1)} \\ 1-\dfrac{|ST_p \cap ST_r|}{|ST_p \cup ST_q \cup ST_r|}, \text{if } N_1 < N_2 \text{ (case 2)} \end{cases} \tag{4.8}$$

In case 1, the expression remains the same. In case 2, a positive quantity $ST_p \cap ST_r$ has been put in place of a negative quantity N_1-N_2. Thus the expression (4.8) reads as

$$\geq \begin{cases} 1-\dfrac{N_1-N_2}{|ST_p \cup ST_r|}, \text{if} N_1 \geq N_2 \\ 1-\dfrac{|ST_p \cap ST_r|}{|ST_p \cup ST_r|}, \text{if} N_1 < N_2 \end{cases}$$
$$\geq \begin{cases} 1-\dfrac{N_1}{|ST_p \cup ST_r|}, \text{if} N_1 \geq N_2 \\ 1-\dfrac{|ST_p \cap ST_r|}{|ST_p \cup ST_r|}, \text{if} N_1 < N_2 \end{cases} \geq \begin{cases} 1-\dfrac{|ST_p \cap ST_r|}{|ST_p \cup ST_r|}, \text{if} N_1 \geq N_2 \\ 1-\dfrac{|ST_p \cap ST_r|}{|ST_p \cup ST_r|}, \text{if} N_1 < N_2 \end{cases} \quad (4.9)$$

where $N_1 = | ST_p \cap ST_q \cap ST_r | \leq | ST_p \cap ST_r |$. Therefore, irrespective of the relationship between N_1 and N_2, $1- PA(a_p, a_q, DB) + 1- PA(a_q, a_r, DB) \geq 1- PA(a_p, a_r, DB)$. Thus, $1- PA(x, y, DB)$ satisfies triangular inequality.

To compute an overall association between two items, we need to express *OA* in terms of supports of frequent itemsets.

Lemma 4.3 *For any two items x, y ∈ I(DB), OA(x, y, DB) can be expressed as follows:*

$$OA(x, y, DB) = \frac{3 \times supp(\{x, y\}, DB) - supp(\{x\}, DB) - supp(\{y\}, DB)}{supp(\{x\}, DB) + supp(\{y\}, DB) - supp(\{x, y\}, DB)} \quad (4.10)$$

Proof $OA(x, y, DB) = PA(x, y, DB) - NA(x, y, DB)$

$$\text{Now, } PA(x, y, DB) = \frac{supp(\{x, y\}, DB)}{supp(\{x\}, DB) + supp(\{y\}, DB) - supp(\{x, y\}, DB)} \quad (4.11)$$

$$\text{Also, } NA(x, y, DB) = \frac{supp(\{x\}, DB) + supp(\{y\}, DB) - 2 \times supp(\{x, y\}, DB)}{supp(\{x\}, DB) + supp(\{y\}, DB) - supp(\{x, y\}, DB)} \quad (4.12)$$

Thus, the lemma follows.

4.4.2 Grouping of Frequent Items

For the purpose of explaining the grouping process, we continue our discussion of Example 4.1.

Example 4.2 Based on *SI*, the forwarded databases are given as follows:

$FD_1 = \{ \{a, e\}, \{b, c, g\}, \{b, e, f\} \}$
$FD_2 = \{ \{ b, c\} \}$
$FD_3 = \{ \{a, b, c\}, \{a, e\} \}$
$FD_4 = \{ \{a, e\}, \{b, c, g\} \}$

Let *size*(*DB*) be the number of transactions in *DB*. Then $size(D_1) = 4$, $size(D_2) = 2$, $size(D_3) = 4$, and $size(D_4) = 2$. The union of all forwarded databases is given as $FD = \{\{a, e\}, \{b, c, g\}, \{b, e, f\}, \{b, c\}, \{a, b, c\}, \{a, e\}, \{a, e\}, \{b, c, g\}\}$.
The transaction $\{a, e\}$ has been shown three times, since it has originated from three data sources. We mine the database *FD* and get the following set of frequent itemsets:

$FIS_1(FD \mid 1/14) = \{ \{a\}\ (4/12), \{b\}\ (5/12) \}$
$FIS_2(FD \mid 1/14) = \{ \{a, b\}\ (1/12), \{a, c\}\ (1/12), \{a, e\}\ (3/12), \{b, c\}\ (4/12), \{b, e\}\ (1/12), \{b, f\}\ (1/12), \{b, g\}\ (2/12) \}$

where $X(\eta)$ denotes the fact that the frequent itemset *X* has support η. All the transactions containing item *x* not belonging to *SI* might not be available in *FD*. Thus other frequent itemsets of size one could not be mined correctly from *FD*. They are not shown in $FIS_1(FD)$. Each frequent itemset extracted from *FD* contains an item from *SI*. The collection of patterns in $FIS_1(FD \mid 1/14)$ and $FIS_2(FD \mid 1/14)$ could be considered as *PB* with reference to Fig. 4.1. Using the frequent itemsets in $FIS_1(FD \mid \alpha)$ and $FIS_2(FD \mid \alpha)$ we might not be able to compute the value of *OA* between two items. The central office of the company requests each branch for the supports of the relevant items (*RI*s) to calculate the overall association between two items. Such information would help the central office to compute exactly the value of the overall association in the union of all databases. Relevant items are the items in $FIS_1(FD \mid \alpha)$ that do not belong to *SI*. In this example, *RI*s are *c*, *e*, *f* and *g*. The supports of relevant items in different databases are given below:

$RI(D_1) = \{ \{c\}\ (1/4), \{e\}\ (2/4), \{f\}\ (1/4), \{g\}\ (2/4) \}$
$RI(D_2) = \{ \{c\}\ (1/2), \{e\}\ (0), \{f\}\ (1/2), \{g\}\ (0) \}$
$RI(D_3) = \{ \{c\}(2/4), \{e\}(1/4), \{f\}(0), \{g\}(1/4) \}$
$RI(D_4) = \{ \{c\}\ (1/2), \{e\}\ (1/2), \{f\}\ (0), \{g\}\ (1/2) \}$

$RI(D_i)$ could be considered as LPB_i with reference to Fig. 4.1, $i = 1, 2, \ldots, n$. We follow here a grouping technique based on the proposed measure of overall association *OA*. If $OA(x, y, D) > 0$ then *y* could be placed in the group of *x*, for $x \in SI = \{a, b\}$, $y \in I(D)$. We explain the procedure of grouping frequent items with the help of following example.

Example 4.3 Here we continue the discussion of Example 4.2. Based on the available supports of local 1-itemsets, we synthesize 1-itemsets in *D* as mentioned in Table 4.6.

Table 4.6 Supports of relevant 1-itemsets in *D*

Itemset ($\{x\}$)	$\{a\}$	$\{b\}$	$\{c\}$	$\{e\}$	$\{f\}$	$\{g\}$
$supp(\{x\}, D)$	4/12	5/12	5/12	4/12	2/12	4/12

We note that the supports of $\{a\}$ and $\{b\}$ are not required to be synthesized, since they could be determined exactly from mining *FD*. The values of *OA* corresponding to itemsets of FIS_2 are presented in Table 4.7.

Table 4.7 Overall association between two items in a frequent 2-itemset in *FD*

Itemset ($\{x, y\}$)	$\{a, b\}$	$\{a, c\}$	$\{a, e\}$	$\{b, c\}$	$\{b, e\}$	$\{b, f\}$	$\{b, g\}$
OA(*x*, *y*, *D*)	−3/4	−3/4	1/5	1/3	−3/4	−2/3	−3/7

In Table 4.7, we find that the items *a* and *e* are positively associated. Thus, item *e* could be placed in the group containing nucleus item *a*. Items *b* and *c* are positively associated as well. Item *c* could be put in the group containing nucleus item *b*. Thus, the output grouping π using the proposed technique comes in the form:

$$\pi(FIS_1(D)|\{a, b\}, 1/12) = \{\text{Group1}, \text{Group2}\},$$

where

$$\begin{aligned}\text{Group 1} &= \{(a, 1.0), (e, 0.2)\}\\ \text{Group 2} &= \{(b, 0.1), (c, 0.33)\}.\end{aligned}$$

Each item in a group is associated with a real number which represents the strength of an overall association between the item and the nucleus item of the group. The proposed grouping technique also constructs the third group of items, i.e., $\{f, g\}$. The proposed study is not concerned with the items in $\{f, g\}$.

Each group grows being centered around a select item. The *i*-th group (G_i) grows centering around the *i*-th select item s_i, $i = 1, 2, \ldots, m$. With respect to group G_i, the item s_i is called the nucleus item of G_i, $i = 1, 2, \ldots, m$. We define a group as follows.

Definition 4.2 The *i*-th group is a collection of frequent items a_j and the nucleus item $s_i \in \text{SI}$ such that $\text{OA}(s_i, a_j, D) > 0, j = 1, 2, \ldots, |G_i|$, and $i = 1, 2, \ldots, m$.

Let us describe the data structures used in the algorithm for finding groups. The set of frequent *k*-itemsets is maintained in an array *FISk*, $k = 1, 2$. After finding *OA* value between two items in a 2-itemset, it is kept in array *IS2*. Thus, the number of itemsets in *IS2* is equal to the number of frequent 2-itemsets extracted from *FD*. A two-dimensional array *Groups* is maintained to store *m* groups. The *i*-the row of *Groups* stores the *i*-th group, for $i = 1, 2, \ldots, m$. The first element of *i*-th row contains the *i*-th select item, for $i = 1, 2, \ldots, m$. In general, the *j*-th element of the *i*-th row contains a pair (*item*, *value*), where *item* refers to the *j*-th item of the *i*-th group and *value* refers to the amount of *OA* value between the i-th select item and *item*, for $j = 1, 2, \ldots, |G_i|$. The grouping algorithm can be outlined as follows.

Algorithm 4.1 Construct m groups of frequent items in D such that i-th group grows being centered around the i-th select item, for $i = 1, 2, \ldots, m$.

procedure *m-grouping* (m, *SI*, N_1, *FIS1*, N_2, *FIS2*, *GSize*, *Groups*)

Input: m, *SI*, N_1, *FIS1*, N_2, *FIS2*
m: the number of select items
SI: set of select items
N_k: number of frequent k-itemsets
FISk: set of frequent k-itemsets
Output: *GSize*, *Groups*
GSize: array of number of elements in each group
Groups: array of m groups

```
01:  for i = 1 to N2 do
02:     IS2(i).value = OA(FIS2(i).item1, FIS2(i).item2, D);
03:     IS2(i).item1 = FIS2(i).item1; IS2(i).item2 = FIS2(i).item2;
04:  end for
05:  for i = 1 to m do
06:     Groups(i)(1).item = SI(i); Groups(i)(1).value = 1.0; GSize(i) = 1;
07:  end for
08:  for i = 1 to N2 do
09:  for j = 1 to m do
10:     if ((IS2(i).item1 = SI(j)) and (IS2(i).value > 0)) then
11:        GSize(j) = GSize(j) + 1; Groups(j)(GSize(j)).item = IS2(i).item2;
12:        Groups(j)(GSize(j)).value = IS2(i).value;
13:     end if
14:     if ((IS2(i).item2 = SI(j)) and (IS2(i).value > 0)) then
15:        GSize(j) = GSize(j) + 1; Groups(j)(GSize(j)).item = IS2(i).item1;
16:        Groups(j)(GSize(j)).value = IS2(i).value;
17:     end if
18:    end for
19:  end for
20:  for i = 1 to m do
21:     sort items of group i in non-increasing order on OA value;
22:  end for
end procedure
```

The algorithm works as follows. Using (4.10), we compute the value of *OA* for each itemset in *FIS2*. After computing *OA* value for a pair of items, we store the items and *OA* value in *IS2*. The algorithm performs these tasks using the for-loop shown in lines 01–04. We initialize each group with the corresponding nucleus item as shown in lines 05–07. A relevant item or an item in *SI* could belong to one or more groups. Thus, we check for the possibility of including each of the relevant items and items in *SI* to each group using the for-loop (lines 09–18). All the relevant items and items in *SI* are covered using for-loop present in lines 08–19. For the purpose of better presentation, we finally sort items of i-th group in non-increasing order on *OA* value, $i = 1, 2, \ldots, m$.

Assume that the frequent itemsets in *FIS1* and *FIS2* are sorted on items in the itemset. Thus, the time complexities for searching an itemset in *FIS1* and *FIS2* are $O(log(N_1))$ and $O(log(N_2))$, respectively. The time complexity of computing present at line 02 is $O(log(N_1))$, since $N_1 > N_2$. The time complexity of calculations carried out in lines 01–04 is $O(N_2 \times log(N_1))$. Lines 05–07 are used to complete all necessary initialization. The time complexity of this program segment is $O(m)$. Lines 08–19 process frequent 2-itemsets and construct m groups. If one of the two items in a frequent 2-itemset is a select item, then other item could be placed in the group of the select item, provided the overall association between them is positive. The time complexity of this program segment is $O(m \times N_2)$. Lines 20–22 present groups in sorted order. Each group is sorted in non-increasing order with respect to the *OA* value. The association of nucleus item with itself is 1.0. Thus the nucleus item is kept at the beginning of the group (line 06). Let the average size of a group be k. Then the time complexity of this program segment is $O(m \times k \times log(k))$. The time complexity of the procedure *m-grouping* is *maximum* $\{ O(N_2 \times log(N_1)), O(m), O(m \times N_2), O(m \times k \times log(k)) \}$, i.e., *maximum* $\{ O(N_2 \times log(N_1)), O(m \times N_2), O(m \times k \times log(k)) \}$.

4.4.3 Experiments

We have carried out several experiments to quantify the effectiveness of the above approach. We present the experimental results using four databases, viz., *retail* (Frequent itemset mining dataset repository 2004), *mushroom* (Frequent itemset mining dataset repository 2004), *T10I4D100K* (Frequent itemset mining dataset repository 2004), and *check*. The database *retail* is real and obtained from an anonymous Belgian retail supermarket store. The database *mushroom* is real and obtained from UCI databases. The database *T10I4D100K* is synthetic and was obtained using a generator from IBM Almaden Quest research group. The database *check* is artificial whose grouping is already known. We have experimented with database *check* to verify that our grouping technique works correctly. We present some characteristics of these databases in Table 4.8. Let *NT*, *AFI*, *ALT*, and *NI* denote the number of transactions, average frequency of an item, average length of a transaction, and number of items in the database, respectively.

Table 4.8 Characteristics of databases used in the experiment

Database	*NT*	*ALT*	*AFI*	*NI*
retail (*R*)	88,162	11.31	99.67	10,000
mushroom (*M*)	8,124	24.00	1,624.80	120
T10I4D100K (*T*)	100,000	11.10	1,276.12	870
check(*C*)	40	3.03	3.10	39

We divide each of these databases into ten databases called here *input databases*. The input databases obtained from R, M, T and C are names as R_i, M_i, T_i, and C_i, respectively, $i = 1, 2, \ldots, 10$. We present some characteristics of the input databases in Tables 4.9 and 4.10.

Table 4.9 Characteristics of input databases obtained from *retail* and *mushroom*

DB	*NT*	*ALT*	*AFI*	*NI*	*DB*	*NT*	*ALT*	*AFI*	*NI*
R_1	9,000	11.24	12.07	8,384	M_1	812	24.00	295.27	66
R_2	9,000	11.21	12.27	8,225	M_2	812	24.00	286.59	68
R_3	9,000	11.34	14.60	6,990	M_3	812	24.00	249.85	78
R_4	9,000	11.49	16.66	6,206	M_4	812	24.00	282.43	69
R_5	9,000	10.96	16.04	6,148	M_5	812	24.00	259.84	75
R_6	9,000	10.86	16.71	5,847	M_6	812	24.00	221.45	88
R_7	9,000	11.20	17.42	5,788	M_7	812	24.00	216.53	90
R_8	9,000	11.16	17.35	5,788	M_8	812	24.00	191.06	102
R_9	9,000	12.00	18.69	5,777	M_9	812	24.00	229.27	85
R_{10}	9,000	11.69	15.35	5,456	M_{10}	816	24.00	227.72	86

Table 4.10 Characteristics of input databases obtained from *T10I4D100K*

DB	*ALT*	*AFI*	*NI*	*DB*	*ALT*	*AFI*	*NI*
T_1	11.06	127.66	866	T_6	11.14	128.63	866
T_2	11.13	128.41	867	T_7	11.11	128.56	864
T_3	11.07	127.65	867	T_8	11.10	128.45	864
T_4	11.12	128.44	866	T_9	11.08	128.56	862
T_5	11.13	128.75	865	T_{10}	11.08	128.11	865

The input databases obtained from database *check* are given as follows:

$C_1 = \{ \{1, 4, 9, 31\}, \{2, 3, 44, 50\}, \{6, 15, 19\}, \{30, 32, 42\} \}$
$C_2 = \{ \{1, 4, 7, 10, 50\}, \{3, 44\}, \{11, 21, 49\}, \{41, 45, 59\} \}$
$C_3 = \{ \{1, 4, 10, 20, 24\}, \{5, 7, 21\}, \{21, 24, 39\}, \{26, 41, 46\} \}$
$C_4 = \{ \{1, 4, 10, 23\}, \{5, 8\}, \{5, 11, 21\}, \{42, 47\} \}$
$C_5 = \{ \{1, 4, 10, 34\}, \{5, 49\}, \{25, 39, 49\}, \{49\} \}$
$C_6 = \{ \{1, 3, 44\}, \{6, 41\}, \{22, 26, 38\}, \{45, 49\} \}$
$C_7 = \{ \{1, 2, 3, 10, 20, 44\}, \{11, 12, 13\}, \{24, 35\}, \{47, 48, 49\} \}$
$C_8 = \{ \{2, 3, 20, 39\}, \{2, 3, 20, 44, 50\}, \{32, 49\}, \{42, 45\} \}$
$C_9 = \{ \{2, 3, 20, 44\}, \{3, 19, 50\}, \{5, 41, 45\}, \{21\} \}$
$C_{10} = \{ \{2, 20, 45\}, \{5, 7, 21\}, \{11, 19\}, \{22, 30, 31\} \}$

In Table 4.11, we present some relevant details regarding different experiments. We have chosen the first 10 frequent items as the select items, except for the last experiment. One could choose select items as the items whose data analyses are needed to be performed.

The first experiment is based on database *retail*. The grouping of frequent items in *retail* is given below:

Table 4.11 Some relevant information regarding experiments

Database	α	*SI*
R	0.03	{0,1,2,3,4,5,6,7,8, 9}
M	0.05	{1, 3, 9, 13, 23, 34, 36, 38, 40, 52}
T	0.01	{2, 25, 52, 240, 274, 368, 448, 538, 561, 630}
C	0.07	{1, 2, 3}

π (*FI*(*retail*) | *SI*, α) = { 0 (1.00); 1 (1.00); 2 (1.00); 3 (1.00); 4 (1.00); 5 (1.00); 6 (1.00); 7 (1.00); 8 (1.00); 9 (1.00) }

Two resulting groups are separated by semicolon (;). The nucleus item in each group is underlined. Each item in a group is associated with a real number shown in bracket. This value represents the strength of the overall association between the item and the nucleus item. The groups are shaded alternately for the purpose of clarity of visualization. We observe that no item in database *retail* is positively associated with the select items using the measure *OA*. This does not necessarily mean that the amount of AE or ME for the experiment is zero. There may exist frequent itemsets of size two such that overall association between two items in each of the itemsets is non-positive and at least one of the two items belongs to the set of select items.

The second experiment is based on database *mushroom*. The grouping of frequent items in *mushroom* is given below:

π (*FI*(*mushroom*) | *SI*, α) = { 1 (1.00), 24 (0.23, 110 (0.12), 29 (0.10), 36 (0.10), 61 (0.10), 38 (0.06), 66 (0.06), 90 (0.01); 3 (1.00); 9 (1.000000); 13 (1.00); 23 (1.00), 93 (0.53), 59 (0.22), 2 (0.14), 39 (0.01), 63 (0.15); 34 (1.00), 86 (0.99), 85 (0.95), 90 (0.80), 36 (0.63), 39 (0.33), 59 (0.23), 63 (0.17), 53 (0.16), 67 (0.13), 24 (0.12), 76 (0.11); 36 (1.00), 85 (0.68), 90 (0.65), 86 (0.63), 34 (0.63), 59 (0.17), 39 (0.16), 63 (0.11), 110 (0.10), 1 (0.10); 38 (1.00), 48 (0.38), 102 (0.19), 58 (0.14), 1 (0.06), 94 (0.05), 110 (0.01); 40 (1.00); 52 (1.00) }

We observe that some frequent items are not included in any of these groups, since their overall associations with each of the select items are non-positive.

The third experiment is based on database *T10I4D100K*. The grouping of frequent items in *T10I4D100K* is given below:

π(*FI*(*T10I4D100K*) | *SI*, α) = { 2 (1.00); 25 (1.00); 52 (1.00); 240 (1.00); 274 (1.00); 368 (1.00); 448 (1.00); 538 (1.00); 561 (1.00); 630 (1.00) }

We observe that databases *retail* and *T10I4D100K* are sparse. Thus, the grouping contains groups of singleton item for these two databases. The overall association between a nucleus item and itself is 1.0. Otherwise, the overall association between a frequent item and a nucleus item is non-positive for these two databases.

The fourth experiment is based on database *check*. The database *check* is constructed artificially to verify the following existing grouping.

$\pi(FI(check) \mid SI, \alpha) = \{$ $(\underline{1}, 1.00), (4, 0.43), (10, 0.43)$; $(\underline{2}, 1.00), (20, 0.43), (3, 0.11)$; $(\underline{3}, 1.00), (44, 0.50), (2, 0.11)$ $\}$.

We have calculated average errors using both trend and proposed approaches. Figures 4.3, 4.4 and 4.5 show the graphs of AE versus the number databases for the first three databases. The proposed model enables us to find actual supports of all the relevant itemsets in a database. Thus, the AE of an experiment for the proposed approach remains 0. As the number of databases increases, the relative presence of a frequent itemset normally decreases. Thus, the error of synthesizing an itemset also increases. Overall, the AE of the experiment using trend approach is likely to increase as the number of databases increases. We observe this phenomenon in Figs. 4.3, 4.4 and 4.5.

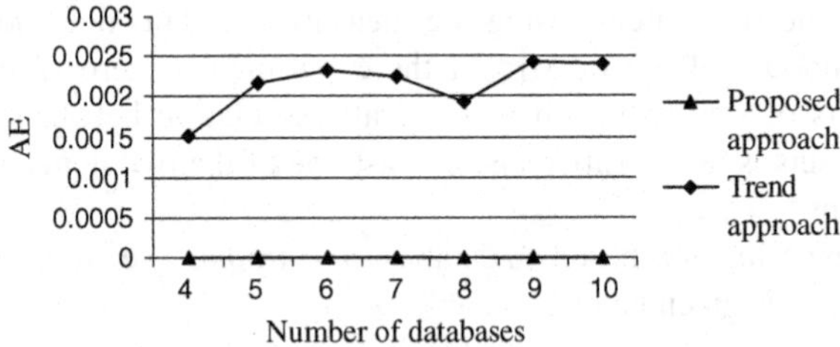

Fig. 4.3 AE vs. the number of the databases from *retail*

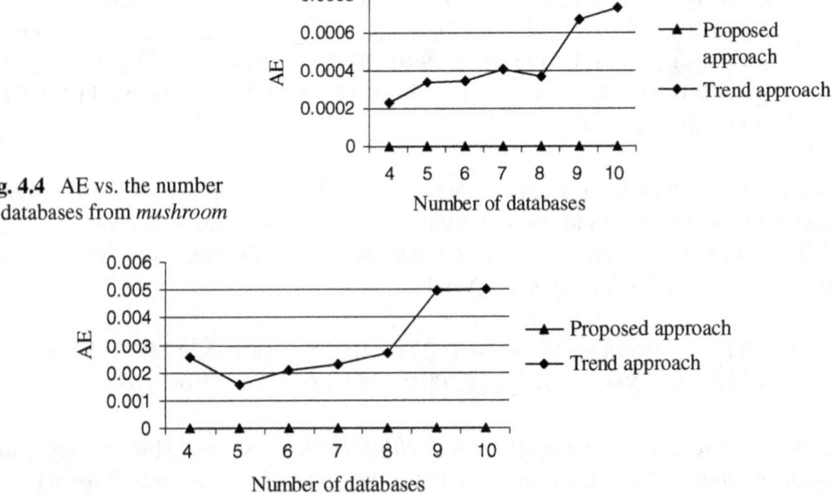

Fig. 4.4 AE vs. the number of databases from *mushroom*

Fig. 4.5 AE vs. the number of databases from *T10I4D100K*

4.5 Related Work

Recently, multi-database mining has been recognized as an important and timely research area in the KDD community. The work reported so far could be classified broadly into two categories: mining/synthesizing patterns in multiple databases and post processing of local patterns. We mention some work related to first category. Wu and Zhang (2003) have proposed a weighting method for synthesizing high-frequency rules in multiple databases. Zhang et al. (2004a) have developed an algorithm to identify global exceptional patterns in multiple databases. When it comes to the second category, Wu et al. (2005) have proposed a technique for classifying multiple databases for multi-database mining. Using local patterns, we have proposed an efficient technique for clustering multiple databases (Adhikari and Rao 2008b).

In the context of estimating support of itemsets in a database, Jaroszewicz and Simovici (2002) have proposed a method using Bonferroni-type inequalities (Galambos and Simonelli 1996). The maximum-entropy approach to support estimation of a general Boolean expression is proposed by Pavlov et al. (2000). But these support estimation techniques are suitable for a single database only.

Zhang et al. (2004b), Zhang (2002) have studied various strategies for mining multiple databases. Proefschrift (2004) has studied data mining on multiple relational databases.

Existing parallel mining techniques (Agrawal and Shafer 1999; Chattratichat et al. 1997; Cheung et al. 1996) could also be used to deal with multi-databases. These techniques might provide expensive solutions for studying select items in multiple databases.

4.6 Conclusions

The proposed measure of overall association *OA* is effective as it considers both positive and negative association between two items. Association analysis of select items in multiple market basket databases is an important as well as highly promising issue, since many data analyses of a multi-branch company are based on select items. One could also apply one of the multi-database mining techniques discussed in Chapter 3. Each technique, except partition algorithm, returns approximate global patterns. On the other hand, the partition algorithm scans each database twice. Therefore, the proposed model of mining global patterns of select items from multiple databases is efficient, since one does not need to estimate the patterns in multiple databases. Moreover, it does not fully scan each database two times.

References

Adhikari A, Rao PR (2007) Study of select items in multiple databases by grouping. In: Proceedings of the 3rd Indian International Conference on Artificial Intelligence, Pune, India, pp. 1699–1718

Adhikari A, Rao PR (2008a) Synthesizing heavy association rules from different real data sources. Pattern Recognition Letters 29(1): 59–71

Adhikari A, Rao PR (2008b) Efficient clustering of databases induced by local patterns. Decision Support Systems 44(4): 925–943

Aggarwal C, Yu P (1998) A new framework for itemset generation. In: Proceedings of the 17th Symposium on Principles of Database Systems, Seattle, WA, pp. 18–24

Agrawal R, Imielinski T, Swami A (1993) Mining association rules between sets of items in large databases. In: Proceedings of ACM SIGMOD Conference, Washington DC, pp. 207–216

Agrawal R, Shafer J (1999) Parallel mining of association rules. IEEE Transactions on Knowledge and Data Engineering 8(6): 962–969

Barte RG (1976) The Elements of Real Analysis. 2nd edition, John Wiley & Sons, New York

Chattratichat J, Darlington J, Ghanem M, Guo Y, Hüning H, Köhler M, Sutiwaraphun J, To HW, Yang D (1997) Large scale data mining: Challenges, and responses. In: Proceedings of the Third International Conference on Knowledge Discovery and Data Mining, pp. 143–146

Cheung D, Ng V, Fu A, Fu Y (1996) Efficient mining of association rules in distributed databases. IEEE Transactions on Knowledge and Data Engineering 8(6): 911–922

Frequent Itemset Mining Dataset Repository (2004) http://fimi.cs.helsinki.fi/data

Galambos J, Simonelli I (1996) Bonferroni-type Inequalities with Applications. Springer, New York

Jaroszewicz S, Simovici DA (2002) Support approximations using Bonferroni-type inequalities. In: Proceedings of Sixth European Conference on Principles of Data Mining and Knowledge Discovery, Helsinki, Finland, pp. 212–223

Klemettinen M, Mannila H, Ronkainen P, Toivonen T, Verkamo A (1994) Finding interesting rules from large sets of discovered association rules. In: Proceedings of the 3rd International Conference on Information and Knowledge Management, Gaithersburg, MD, pp. 401–407

Liu B, Hsu W, Ma Y (1999) Pruning and summarizing the discovered associations. In: Proceedings of the 5th International Conference on Knowledge Discovery and Data Mining, San Diego, CA, pp. 125–134

Pavlov D, Mannila H, Smyth P (2000) Probabilistics models for query approximation with large sparse binary data sets. In: Proceedings of Sixteenth Conference on Uncertainty in Artificial Intelligence, San Francisco, CA, pp. 465–472

Proefschrift (2004) Multi-relational data mining, Ph D thesis, Dutch Graduate School for Information and Knowledge Systems, Aan de Universiteit Utrecht

Pyle D (1999) Data Preparation for Data Mining. Morgan Kufmann, San Francisco

Silberschatz A, Tuzhilin A (1996) What makes patterns interesting in knowledge discovery systems. IEEE Transactions on Knowledge and Data Engineering 8(6): 970–974

Silverstein C, Brin S, Motwani R (1998) Beyond market baskets: Generalizing association rules to dependence rules. Data Mining and Knowledge Discovery 2(1): 39–68

Tan P-N, Kumar V, Srivastava J (2002) Selecting the right interestingness measure for association patterns. In: Proceedings of SIGKDD Conference, Alberta, Canada, pp. 32–41

Wu X, Zhang S (2003) Synthesizing high-frequency rules from different data sources. IEEE Transactions on Knowledge and Data Engineering 14(2): 353–367

Wu X, Zhang C, Zhang S (2005) Database classification for multi-database mining. Information Systems 30(1): 71–88

Xin D, Han J, Yan X, Cheng H (2005) Mining compressed frequent-pattern sets. In: Proceedings of the 31st VLDB Conference, Trondheim, Norway, pp. 709–720

Zhang S (2002) Knowledge discovery in multi-databases by analyzing local instances, Ph D thesis, Deakin University

Zhang C, Liu M, Nie W, Zhang S (2004a) Identifying global exceptional patterns in multi-database mining. IEEE Computational Intelligence Bulletin 3(1): 19–24

Zhang S, Wu X, Zhang C (2003) Multi-database mining. IEEE Computational Intelligence Bulletin 2(1): 5–13

Zhang S, Zhang C, Wu X (2004b) Knowledge Discovery in Multiple Databases. Springer, Berlin

Chapter 5
Enhancing Quality of Knowledge Synthesized from Multi-database Mining

Multi-database mining using local pattern analysis could be considered as an approximate method of mining multiple large databases. Assuming this point of view, it might be required to enhance the quality of knowledge synthesized from multiple databases. Also, many decision-making applications are directly based on the available local patterns present in different databases. The quality of synthesized knowledge/decision based on local patterns present in different databases could be enhanced by incorporating more local patterns in the knowledge synthesizing/processing activities. Thus, the available local patterns play a crucial role in building efficient multi-database mining applications. We represent patterns in a condensed form by employing a so-called ACP (antecedent-consequent pair) coding. It allows one to consider more local patterns by lowering further the user-defined characteristics of discovered patterns, like minimum support and minimum confidence. The ACP coding enables more local patterns participate in the knowledge synthesizing/processing activities and thus the quality of synthesized knowledge based on local patterns becomes enhanced significantly with regard to the synthesizing algorithm and required computing resources. To secure a convenient access to association rule, we introduce an index structure. We demonstrate that ACP coding represents rulebases by making use of the least amount of storage space in comparison to any other rulebase representation technique. Furthermore we present a technique for storing rulebases in the secondary storage.

5.1 Introduction

In Chapters 2, 3, and 4, we have discussed how to improve multi-database mining by adopting different mining techniques. Also, we have learnt that a single multi-database mining technique might not be sufficient in all situations. Chapters 2 and 3 present different variations of multi-database mining using local pattern analysis. Multi-database mining using local pattern analysis could be considered as an approximate method of mining multiple large databases. In this chapter, we employ a coding, referred to as antecedent-consequent pair (ACP) coding, to improve the quality of synthesized knowledge coming from multi-database mining. The ACP coding enables an efficient storage for association rules in multiple databases space.

A. Adhikari et al., *Developing Multi-database Mining Applications*, Advanced Information and Knowledge Processing, DOI 10.1007/978-1-84996-044-1_5,

One could extract knowledge of better quality by storing more association rules in the main memory. In this way, applications dealing with association rules in multiple databases become more efficient.

Consider a multi-branch company that operates at different locations. Each branch generates a large database and subsequently we have to deal with multiple large databases. In particular, the company might be interested in identifying the global association rules in the union of all databases. Let $X \rightarrow Y$ be an association rule extracted from a few databases. Then local pattern analysis might return approximate association rule $X \rightarrow Y$ in the union of all databases, since the association rule might not get extracted from all the databases. As the higher number of data sources report the association rule, the quality of synthesized association rule gets elevated. We discuss how to enhance the quality of synthesized association rules in multiple databases.

Many multi-database mining applications often handle a large number of patterns. In multi-database mining applications, local patterns could be used in two ways. In the first category of applications, global patterns are synthesized from local patterns (Wu and Zhang 2003; Zhang et al. 2004). Synthesized global patterns could be used in various decision-making problems. In the second category of applications, various decisions are taken based on the local patterns present in different databases (Adhikari and Rao 2008; Wu et al. 2005). Thus, the available local patterns could play an important role in finding a solution to a problem. For a problem positioned in the first category, the quality of a global pattern is influenced by the pattern synthesizing algorithm and the locally available patterns. Also, we observe that a global pattern synthesized from local patterns might be approximate. For a given pattern synthesizing algorithm, one could enhance the quality of synthesized patterns by increasing the number of local patterns in a process of knowledge synthesis. For the problems pertinent to the second category, the quality of the resulting decision is implied by the quality of measure used in the decision-making process. Again, the quality of measure is based on the correctness of the measure itself and the available local patterns. For the purpose of database clustering, Wu et al. (2005) have proposed two such measures expressing similarity between two databases. For a given measure of decision-making, one could enhance the quality of decision by increasing the number of local patterns in the decision making process. In other words, the number of available local patterns plays a crucial role in building efficient multi-database mining applications. One could increase the number of local patterns by lowering the user-defined inputs, such as minimum support and minimum confidence. More patterns could be stored in main memory by applying a space efficient pattern base representation technique. In this chapter, we present the ACP coding (Adhikari and Rao 2007) to represent a set of association rules present in different databases space.

As before, let D_i be the database corresponding to the i-th branch of the company, $i = 1, 2, \ldots, n$, while D stands for the union of these databases. The data mining model adopted in this chapter for association rule is the support (supp)-confidence (conf) framework established by Agrawal et al. (1993). The set of association rules extracted from a database is called a *rulebase*. Before proceeding with the algorithmic details, let us introduce some useful notations. Let RB_i be the rulebase corresponding to database D_i at the *minimum support level* α and *minimum confidence level* β, $i = 1, 2, \ldots, n$. Also, let *RB* be the union of rulebases corresponding to different databases. Many interesting algorithms have been reported on mining association rules in a database (Agrawal and Srikant 1994; Han et al. 2000; Savasere et al. 1995). Let *T* be a technique for representing *RB* in main memory. Let φ and ψ denote the pattern synthesizing algorithm and computing resource used for a data mining application, respectively. Also, let $\xi(RB \mid T, \alpha, \beta, \varphi, \psi)$ denote the collection of synthesized patterns over *RB* at a given tuple $(T, \alpha, \beta, \varphi, \psi)$. The quality of synthesized patterns could be enhanced if the number of local patterns increases. Thus, the quality of $\xi(RB \mid T, \alpha_1, \beta_1, \varphi, \psi)$ is lower than the quality of $\xi(RB \mid T, \alpha_2, \beta_2, \varphi, \psi)$, if $\alpha_2 < \alpha_1$ and $\beta_2 < \beta_1$. Thus, the problem of enhancing the quality of synthesized patterns translates to the problem of designing a space-efficient technique for representing rulebases corresponding to different databases.

As the frequent itemsets are the natural form of compression for association rules, the following reasons motivate us to compress association rules rather than frequent itemsets. Firstly, applications dealing with the association rules could be developed efficiently. Secondly, a frequent itemset might not generate any association rule at a given minimum confidence.

In this chapter, we present a space efficient technique to represent *RB* in a main memory. Let $SP^T(RB \mid \alpha, \beta, \psi)$ and $SP^T_{min}(RB \mid \alpha, \beta, \psi)$ describe the amount of space (expressed in bits) and minimum amount of space (expressed in bits) consumed by *RB* using a rulebase representation technique *T*, respectively. We observe that a rulebase representation technique might not represent *RB* at its minimum level because of the random nature of the set of transactions contained in the database. In other words, a frequent itemset might not generate all the association rules. For example, the association rule $X \rightarrow Y$ might not get extracted from any one of the given databases, even if the itemset $\{X, Y\}$ is frequent in some databases. Thus $SP^T_{min}(RB \mid \alpha, \beta, \psi) \leq SP^T(RB \mid \alpha, \beta, \psi)$, for a given tuple (α, β, ψ), where $0 < \alpha \leq \beta \leq 1$. Let Γ be the set of all techniques for representing a set of association rules. We are interested in finding a technique $T_1 \in \Gamma$ for representing *RB*, such that $SP^{T_1}(RB \mid \alpha, \beta, \psi) \leq SP^T(RB \mid \alpha, \beta, \psi)$, for all $T \in \Gamma$. Let $SP_{min}(RB \mid \alpha, \beta, \psi) =$ *minimum* $\{ SP^{\mathrm{T}}_{min}(RB \mid \alpha, \beta, \psi)$: $T \in \Gamma\}$. The efficiency of *T* for representing *RB* is evaluated by comparing $SP^T(RB \mid \alpha, \beta, \psi)$ with $SP_{min}(RB \mid \alpha, \beta, \psi)$. We would like to design an efficient rulebase representation technique T_1 such that $SP^{T_1}(RB \mid \alpha, \beta, \psi) \leq SP^T(RB \mid \alpha, \beta, \psi)$, for $T \in \Gamma$.

The study presented in this chapter is based on a collection of rulebases RB_i, $i = 1, 2, \ldots, n$. One could lower α and β further so that each RB_i represents the corresponding database reasonably well. The work is not concerned with mining branch databases. The coding presented in this chapter reduces *RB* significantly, so that the coded *RB* becomes available in the main memory during the execution of pattern processing/synthesizing algorithm. The benefits of coding *RB* are given as follows. Firstly, the quality of processed/synthesized knowledge gets enhanced, since the number of local association rules participate in the pattern processing/synthesizing algorithm is higher. Secondly, the pattern processing/synthesizing algorithm could access all the local association rules conveniently, since coded *RB* becomes available in the main memory. This arrangement might be possible, since the coded *RB* is reasonably small. For the purpose of achieving latter benefit, we present an index structure to access the coded association rules. Finally, the coded RB and the corresponding index table could be stored in the secondary storage for the usage of different multi-database mining applications. The following issues are discussed:

- We present the ACP coding, for representing rulebases corresponding to different databases space efficiently. It enables us to incorporate more association rules for synthesizing global patterns or decision-making activities.
- We present an index structure to access the coded association rules.
- We prove that the ACP coding represents *RB* using the least amount of storage space in comparison to any other rulebase representation technique.
- We present a technique for storing rulebases corresponding to different databases in the secondary storage.
- We conduct experiments to express the effectiveness of the proposed approach.

The chapter is organized as follows. In Section 5.2, we discuss related work. A simple coding, called SBV coding, for representing different rulebases is presented in Section 5.3. In Section 5.4, we present the ACP coding for representing rulebases space. Experimental results are covered in Section 5.5.

5.2 Related Work

Our objective is to enhance the quality of decisions induced by local association rules. To achieve this objective, we present ACP coding for reducing the storage space of rulebases corresponding to different databases. There are three approaches to reducing the amount of storage space of different rulebases. Firstly, one could devise a mining technique for reducing the number of association rules extracted from a database. Secondly, one could adopt a suitable data structure for reducing the storage space for representing association rules in main memory. Thirdly, one could devise a post-mining technique along with a suitable data structure for reducing storage space required for association rules in the main memory. The first and second approaches to reducing the storage space are normally followed during a data mining task. Here, we concentrate on the third approach to reduce the storage space of different rulebases.

While mining association rules, we observe that there may exist many redundant association rules in a database. Using the semantics based on the closure of the Galois connection (Fraleigh 1982), one could define a condensed representation of association rules (Pasquier et al. 2005). This representation is characterised by frequent closed itemsets and their generators (Zaki and Ogihara 1998). It contains the non-redundant association rules having minimal antecedent and maximal consequent. These rules are the most relevant since they are the most general non-redundant association rules. Mining association rule is iterative and interactive nature. The user has to refine his/her mining queries until he/she is satisfied with the discovered patterns. To support such an interactive process, an optimized sequence of queries is proposed by means of a cache that stores information from previous queries (Jeudy and Boulicaut 2002). The technique uses condensed representations like free and closed itemsets for both data mining and caching. A condensed representation of the frequent patterns called disjunction-free sets (Bykowski and Rigotti 2003), could be used to regenerate all the frequent patterns and their exact frequencies without any access to the original data. In what follows, we discuss work related to the second approach to reducing the storage space of different rulebases.

Shenoy et al. (2000) have proposed a vertical mining algorithm that applies some optimization techniques for mining frequent itemsets in a database. Coenen et al. (2004) have proposed two new structures for association rule mining, the so-called T-tree, and P-tree, together with associated algorithms. The T-tree offers significant advantages in terms of generation time, and storage requirements compared to hash tree structures (Agrawal and Srikant 1994). The P-tree offers significant pre-processing advantages in terms of generation time and storage requirements compared to the FP-tree (Han et al. 2000). The T-tree and P-tree data structures are useful during the mining of a database. At the top level, T-tree stores supports for 1-itemsets, the second level for 2-itemsets, and so on. In T-tree, each node is an object containing support and a reference to an array of child T-tree nodes. The implementation of this data structure could be optimised by storing levels in the tree in the form of arrays, thus reducing the number of links needed and providing indexing. P-tree is different from T-tree in some ways. The idea behind the construction of P-tree can be outlined as follows. At the first pass of scanning input data, the entire database is copied into a data structure, which maintains all the relevant aspects of the input, and then mines this structure. P-tree offers two advantages: (i) it merges the duplicated records and records with common leading substrings, thus reducing the storage and processing requirements, and (ii) it allows partial counts of the support for individual nodes within the tree to be accumulated effectively as the tree is constructed. The top level is comprised of an array of nodes, each index describing a 1-itemset, with child references to body P-tree nodes. Each node at the top level contains the following fields: (i) a field for the support, and (ii) a link to a body P-tree node. A body P-tree node contains the following fields: (i) a support field, (ii) an array of short integers for the itemset that the node represents, and (iii) child and sibling links to further P-tree nodes. T-tree and P-tree structures are not suitable for storing and accessing association rules. These structures do not provide explicit provisions for storing confidence and database identification of association rules in different databases. It is difficult to handle effectively association rules in different databases during post-mining of rulebases corresponding to different databases. Ananthanarayana et al. (2003) have proposed PC-tree to represent data completely and minimally in main memory. It is built by scanning database only once. It could be used to represent dynamic databases with the help of knowledge that is either static or dynamic. It is not suitable for storing and accessing association rules. Furthermore PC-tree lacks the capability of handling association rules in different databases during post-mining of rulebases corresponding to different databases.

The proposed work falls under the third category of solutions to reducing storage of different rulebases. It is useful for handling association rules effectively during post-mining of association rules in different databases. No work has been reported so far under this category.

In the context of mining good quality of knowledge from different data sources, Su et al. (2006) have proposed a framework for identifying trustworthy knowledge from external data sources. Such framework might not be useful in this context.

Zhang and Zaki (2002) have edited a study on various problems related to multi-database mining. Zhang (2002) studied various strategies for mining multiple databases. Kum et al. (2006) have presented an algorithm, ApproxMAP, to mine approximate sequential patterns, called *consensus patterns*, from large sequence databases in two steps. First, sequences are clustered by similarity. Then, consensus patterns are mined directly from each cluster through multiple alignments.

5.3 Simple Bit Vector (SBV) Coding

We need to process all the association rules in different local databases for synthesizing patterns, or decision-making applications. We use a tuple (*ant*, *con*, *s*, *c*) to represent an association rule in a symbolic manner, where *ant*, *con*, *s*, and *c* represent antecedent, consequent, support and confidence of the association rule *ant* → *con*, respectively. The following example serves as a pertinent illustration of this representation.

Example 5.1 A multi-branch company has four branches. Let $\alpha = 0.35$, and $\beta = 0.45$. The rulebases corresponding to these different databases are given below:

RB_1 = { (*A*, *C*, 1.0, 1.0), (*C*, *A*, 1.0, 1.0), (*A*, *B*, 0.42, 0.42), (*B*, *A*, 0.42, 0.74), (*B*, *C*, 0.40, 0.71), (*C*, *B*, 0.40, 0.40), (*A*, *BC*, 0.36, 0. 36), (*B*, *AC*, 0.36, 0.64), (*C*, *AB*, 0.36, 0.36), (*AB*, *C*, 0.36, 0.74), (*AC*, *B*, 0.36, 0.36), (*BC*, *A*, 0.36, 0.90) }

RB_2 = { (*A*, *C*, 0.67, 0.67), (*C*, *A*, 0.67, 1.0) }

RB_3 = { (*A*, *C*, 0.67, 0.67), (*C*, *A*, 0.67, 1.0), (*A*, *E*, 0.67, 0.67), (*E*, *A*, 0.67, 1.0) }

RB_4 = { (*F*, *D*, 0.75, 0.75), (*D*, *F*, 0.75, 1.0), (*F*, *E*, 0.50, 0.50), (*E*, *F*, 0.50, 1.0), (*F*, *H*, 0.50, 0.50), (*H*, *F*, 0.50, 1.0) }.

One could conveniently represent an association rule using an object (or a record). A typical object representing an association rule consists of following attributes: database identification, number of items in the antecedent, items in the antecedent, number of items in the consequent, items in the consequent, support and confidence. We further calculate the space requirement for such an object by continuing Example 5.1.

Example 5.2 A typical compiler represents an integer and a real number using 4 bytes and 8 bytes, respectively. An item could be considered as an integer. Consider the association rule (*A, BC,* 0.36, 0.36) of RB_1. Each of the following components of an association rule could consume 4 bytes: database identification, number of items in the antecedent, item A, number of items in the consequent, item *B*, and item *C*. Support and confidence of an association rule could consume 8 bytes each. The association rule (*A, BC,* 0.36, 0.36) of RB_1 thus consumes 40 bytes. The association rule (*A, C,* 1.0, 1.0) of RB_1 could consume 36 bytes. Thus, the amount of space required to store four rulebases is equal to ($18 \times 36 + 6 \times 40$) bytes, i.e. 7,104 bits. A technique without optimization (TWO) could consume 7,104 bits to represent these rulebases.

Let *I* be the set of all items in *D*. Let *X*, *Y* and *Z* be three itemsets such that $Y, Z \subseteq X$. Then $\{Y, Z\}$ forms a *2-itemset partition* of *X* if $Y \cup Z = X$, and $Y \cap Z = \phi$. We define *size* of itemset *X* as the number of items in *X*, denoted by $|X|$. Then, we have $2^{|X|}$ 2-itemset partitions of *X*. For example, $\{\{a\}, \{b, c\}\}$ is a 2-itemset partition of $\{a, b, c\}$. An association rule $Y \rightarrow Z$ corresponds to a 2-itemset partition of *X*, for $Y, Z \subseteq X$. The antecedent and consequent of an association rule are non-null. We arrive at the following lemma.

Lemma 5.1 *An itemset X can generate maximum* $2^{|X|} - 2$ *association rules for* $|X| \geq 2$.

Let there are 10 items. The number of itemsets using 10 items is 2^{10}. Thus, 10 bits would be enough to represent an itemset. The itemset *ABC*, i.e. {*A*, *B*, *C*} could be represented by the bit combination 1110000000. 2-itemset partitions of *ABC* are $\{\phi, ABC\}$, $\{A, BC\}$, $\{B, AC\}$, $\{C, AB\}$, $\{AB, C\}$, $\{AC, B\}$, $\{BC, A\}$, and $\{ABC, \phi\}$. Number of 2-itemset partitions of a set containing 3 items is 2^3. Every 2-itemset partition corresponds to an association rule, except the partitions $\{\phi, ABC\}$ and $\{ABC, \phi\}$. For example, the partition {*A*, *BC*} corresponds to the association rule $A \rightarrow BC$. Thus, 3 bits are sufficient to identify an association rule generated from *ABC*. If the number of items is large, then this method might take significant amount of memory space to represent itemsets and the association rules generated from the itemsets. Thus, this technique is not suitable to represent association rules in databases containing large number of items.

5.3.1 Dealing with Databases Containing Large Number of Items

We explain the crux of the SBV coding with the help of the following example.

Example 5.3 We continue here the discussion we started in Example 5.1. Let the number of items be 10,000. We need 14 bits to identify an item, since $2^{13} < 10{,}000 \leq 2^{14}$. We assume that the support and confidence of an association rule are represented using 5 decimal digits. Thus support/confidence value 1.0 could be represented as 0.99999. We use 17-bit binary number to represent support/confidence, since $2^{16} < 99{,}999 \leq 2^{17}$.

Let us consider the association rule $(A, BC, 0.36107, 0.36107)$ of RB_1. There are 4 databases viz., D_1, D_2, D_3, and D_4. We need 2 bits to identify a database, since $2^1 < 4 \leq 2^2$. Also 4 bits are enough to represent the number of items in an association rule. We place bit 1 at the beginning of binary representation of an item, if it appears in the antecedent of the association rule. We use bit 0 at the beginning of binary representation of an item, if it appears in the consequent of the association rule. Using this arrangement, the lengths of the antecedent and consequent are not required to be stored. The following bit vector could represent the above association rule

00001110000000000000010000000000000010

1 2 3 4 5 6

000000000000001101000110100001011 01000110100001011

7 8 9 10

The components of above bit vector are explained below.
Component 1 represents the first database (i.e., D_1)
Component 2 represents the number of items in the association rule (i.e., 3)
Component 3 (i.e., bit 1) implies that the current item (i.e., item A) belongs to antecedent
Component 4 represents item A (i.e., item number 1)
Component 5 (i.e., bit 0) implies that the current item (i.e., item B) belongs to consequent
Component 6 represents item B (i.e., item number 2)
Component 7 (i.e., bit 0) implies that the current item (i.e., item C) belongs to consequent
Component 8 represents item C (i.e., item number 3)
Component 9 represents support of association rule
Component 10 represents confidence of association rule

The storage space required for an association rule containing two items and three items are 70 and 85 bits, respectively. Therefore, the amount of storage space required to represent different rulebases is equal to $(18 \times 70 + 6 \times 85)$ bits, i.e., 1,770 bits. A technique without optimization could consume 7,104 bits (as mentioned in Example 5.2) to represent the same structure. We note that SBV coding significantly reduces the amount of storage space for representing different rulebases.

In the following section, we consider a special case of bit vector coding. It optimizes the storage space for representing different rulebases which is based on the fact that many association rules have the same antecedent-consequent pair. Before we move on to the next section, we consider the following lemma.

Lemma 5.2 *Let there are p items. Let m be the minimum number of bits required to represent an item. Then,* $m = \lceil \log_2 (p) \rceil$.

Proof We have $2^{m-1} < p \leq 2^m$, for an integer m. Thus we get $m < \log_2 (p) + 1$, and $\log_2 (p) \leq m$, since $\log_k(x)$ is a monotonically increasing function of x, for $k > 1$. Combining these two inequalities we obtain $\log_2 (p) \leq m < \log_2 (p) + 1$.

5.4 Antecedent-Consequent Pair (ACP) Coding

The central office generates sets of frequent itemsets from different rulebases. Let FIS_i be the set of frequent itemsets generated from RB_i, $i = 1, 2, \ldots, n$. Also let *FIS* denote the union of all frequent itemsets being reported from different databases. In a symbolic way, we denote a frequent itemset as a pair (itemset, support). The association rules (*F*, *D*, 0.75, 0.75) and (*D*, *F*, 0.75, 1.0) of RB_4 generate the following frequent itemsets: (*D*, 0.75), (*F*, 1.0) and (*DF*, 0.75). In the following example, we generate FIS_i, for $i = 1, 2, \ldots, n$.

Example 5.4 Continuing Example 5.1, the sets of frequent itemsets generated by the central office comes as follows:

FIS_1 = { (*A*, 1.0), (*C*, 1.0), (*B*, 0.57), (*AC*, 1.0), (*AB*, 0.42), (*BC*, 0.40), (*ABC*, 0.36) }
FIS_2 = { (*A*, 1.0), (*C*, 0.67), (*AC*, 0.67) }
FIS_3 = { (*A*, 1.0), (*C*, 0.67), (*E*, 0.67), (*AC*, 0.67), (*AE*, 0.67) }
FIS_4 = { (*D*, 0.75), (*E*, 0.50), (*F*, 1.0), (*H*, 0.50), (*DF*, 0.75), (*EF*, 0.50), (*FH*, 0.50) }.

The ACP coding is a special case of bit vector coding, where antecedent-consequent pairs of the associations rules are coded in a specific order. The ACP coding is lossless (Sayood 2000) and similar to the Huffman coding (Huffman 1952). The ACP coding and the Huffman coding are not the same, in the sense that an ACP code may be a prefix of another ACP code. Then a question arises: how does a search procedure detect antecedent-consequent pair of an association rule correctly? We arrive at the answer to this question in Section 5.4.1.

Let *X* be a frequent itemset generated from an association rule. Also, let $f(X)$ be the number of rulebases that generate itemset *X*. Furthermore let $f_i(X) = 1$, if *X* is extracted from the *i*-th database, and $f_i(X) = 0$, otherwise; for $i = 1, 2, \ldots, n$. Then, $f(X) \leq \sum_{i=1}^{n} f_i(X)$. The central office sorts the frequent itemsets *X* using $|X|$ as the primary key and $f(X)$ as the secondary key, for $X \in FIS$ and $|X| \geq 2$. Initially, the itemsets are sorted on size in non-decreasing order. Then the itemsets of the same size are sorted on $f(X)$ in non-increasing order. If $f(X)$ is high then the number of association rules generated from *X* is expected to be high. Therefore, we represent antecedent-consequent pair of such an association rule using a code of smaller size. We explain the essence of this coding with the help of Example 5.5.

Example 5.5 We continue here the discussion of Example 5.4. We sort all the frequent itemsets of size greater than or equal to 2. Sorted frequent itemsets are presented in Table 5.1.

Table 5.1 Sorted frequent itemsets of size greater than or equal to 2

X	*AC*	*AB*	*AE*	*BC*	*DF*	*EF*	*FH*	*ABC*
$f(X)$	3	1	1	1	1	1	1	1

The coding process is described as follows. Find an itemset that has a maximal *f*-value. Itemset *AC* has the maximum *f*-value. We code *AC* as 0. The maximum number of association rules could be generated from *AC* is two. Thus we code association rules $A \rightarrow C$ and $C \rightarrow A$ as 0 and 1, respectively. Now, 1-digit codes are no more available. Then we find an itemset that has a second maximal *f*-value. We choose *AB*. We could have chosen any itemset from {*AB*, *AE*, *BC*, *DF*, *EF*, *FH*}, since every itemset in the set has the same size and the same *f*-value. We code *AB* as 00. The maximum number of association rules could be generated from *AB* is two. Thus we code the association rules $A \rightarrow B$ and $B \rightarrow A$ as 00 and 01, respectively. We follow in the same way and code the association rules $A \rightarrow E$ and $E \rightarrow A$ as 10 and 11, respectively. Now, we have constructed 2-digit codes. Finally, we choose *ABC*. We code *ABC* as 0000. The association rules $A \rightarrow BC$, $B \rightarrow AC$, $C \rightarrow AB$, $AB \rightarrow C$, $AC \rightarrow B$, and $BC \rightarrow A$ get coded as 0000, 0001, 0010, 0011, 0100, and 0101, respectively. Each frequent itemset receives a code. We call it an *itemset code*. Also, antecedent-consequent pair of an association rule is assigned a code. We call it a *rule code*.

Now an association rule could be represented in the main memory using the following components: database identification number, ACP code, support and confidence. Let n be the number of databases. Then we have $2^{k-1} < n \leq 2^k$, for an integer k. Thus, we need k bits to represent the database identification number. We represent support/confidence using p decimal digits. If we represent a fraction f using an integer d while f is given through the formula: $f = d \times 10^{-p}$. We represent support/confidence by storing the corresponding integer. The following lemma determines the minimum number of binary digits required to store a decimal number.

Lemma 5.3 *A p-digit decimal number can be represented by a* $\lceil p \times \log_2 10 \rceil$*-digit binary number.*

Proof Let t be the minimum number of binary digits required to represent a p-digit decimal number x. Then we have $x < 10^p < 2^t$. So, $t > p \times \log_2 10$, since $\log_k(y)$ is a monotonically increasing function of y, for $k > 1$. Thus the minimum integer t for which $x < 2^t$ is true is given as $\lceil p \times \log_2 10 \rceil$.

The following lemma specifies the minimum amount of storage space required to represent *RB* under some conditions.

Lemma 5.4 *Let M be the number of association rules having distinct antecedent-consequent pairs among N association rules extracted from n databases, where* $2^{m-1} < M \leq 2^m$, *and* $2^{p-1} < n \leq 2^p$, *for some positive integers m and p. Suppose the support and confidence of an association rule are represented by a fractions containing k digits after the decimal point. Assume that a frequent itemset X generates all possible associationrules, for* $X \in FIS$, *and* $|X| \geq 2$. *Then the minimum amount of storage space required to represent RB in the main memory is given as follows.*

$$SP^{\text{ACP coding}}_{\text{min, main}}(RB|\alpha,\ \beta,\ \psi) = M \times (m-1) - 2 \times (2^{m-1} - m)$$
$$+N \times (p + 2 \times \lceil k \times log_2 10 \rceil)\ bits,$$

if $M < 2^m - 2$; *and*

$$SP^{\text{ACP coding}}_{\text{min, main}}(RB|\alpha,\ \beta,\ \psi) = M \times m - 2 \times (2^m - m - 1)$$
$$+N \times (p + 2 \times \lceil k \times log_2 10 \rceil)\ bits,\ otherwise.$$

Proof p bits are required to identify a database. The amount of memory required to represent database identifiers of N association rules is equal to $P = N \times p$ bits. The minimum amount of memory required to represent both support and confidence of N association rules is equal to $Q = N \times 2 \times \lceil k \times \log_2 10 \rceil$ bits (as shown in Lemma 5.3). Let R be the minimum amount of memory required to represent ACPs of M association rules. The expression R could be obtained from the fact that 2^1 ACPs are of length 1, 2^2 ACPs are of length 2, and so on. The expression of R is given as follows.

$$R = \sum_{i=1}^{m-2} i \times 2^i + (m-1) \times \left(M - \sum_{i=1}^{m-2} 2^i\right) \text{ bits, if } \left(M - \sum_{i=1}^{m-2} 2^i\right) < 2^{m-1}\text{; and}$$
$$R = \sum_{i=1}^{m-1} i \times 2^i + m \times \left(M - \sum_{i=1}^{m-1} 2^i\right) \text{ bits, if } \left(M - \sum_{i=1}^{-2} 2^i\right) \geq 2^{m-1}. \tag{5.1}$$

R assumes second form of expression for a few cases. For example, if ($M = 15$) then R assumes second form of expression. The ACP codes are given as follows: 0, 1, 00, 01, 10, 11, 000, 001, 010, 011, 100, 101, 110, 111, 0000. Then, the minimum amount of storage space required to represent RB is equal to ($P + Q + R$) bits. Now, $\sum_{i=1}^{m-2} 2^i = 2^{m-1} - 2$, and $\sum_{i=1}^{m-2} i \times 2^i = (m-3) \times 2^{m-1} + 2$. Thus the lemma follows.

In the following example, we calculate the amount of storage space for representing rulebases of Example 5.1.

Example 5.6 The discussion of Example 5.1 is continued here. The number of association rules in RB is 24. With reference to Lemma 5.4, we have $N = 24$, $M = 20$, and $n = 4$. Thus, $m = 5$, and $p = 2$. Assume that the support and confidence of an association rule are represented by fractions containing 5 decimal digits.

Thus $k = 5$. Then, the minimum amount of storage space required to represent RB is 922 bits.

The ACP coding may assign some codes for which there exists no associated rule. Let ABC be a frequent itemset extracted from some databases. Assume that the association rule $AC \rightarrow B$ is not extracted from any database that extracts ABC. Let the itemset code corresponding to ABC is 0000. Then the ACP code for $AC \rightarrow B$ is 0100, i.e., the 4-th association rule generated from ABC. Therefore the ACP coding does not always store rulebases at the minimum level.

All rule codes are the ACP codes. But, the converse statement is not true. Some ACP codes do not have assigned association rules, since the assigned association rules are not extracted from any one of the given databases. An ACP code X is *empty* if X is not a rule code.

Lemma 5.5 *Let $X \in FIS$ such that $|X| \geq 2$. We assume that X generates at least one association rule. Let m (≥ 2) be the maximum size of a frequent itemset in FIS. Let n_i be the number of distinct frequent itemsets in FIS of size i, $i = 2, 3, \ldots, m$. Then the maximum number of empty ACP codes is equal to $\sum_{i=2}^{m}(2^i - 3) \times n_i$.*

Proof In the extreme case, only one association rule is generated for each frequent itemset X in *FIS*, such that $|X| \geq 2$. Using Lemma 5.1, one can note that a frequent itemset X could generate maximum $2^{|X|}-2$ association rules. In such a situation, $2^{|X|}-3$ ACP codes are empty for X. Thus, the maximum number of empty ACP codes for the frequent itemsets of size i is equal to $(2^i-3) \times n_i$. Hence the result follows.

To search an association rule we maintain all the itemsets in an index table along with their itemset codes such that the size of an itemset is greater than one. We generate rule codes of the association rules from the corresponding itemset code. In Section 5.4.1, we discuss a procedure for constructing index table and accessing mechanism for the association rules.

5.4.1 Indexing Rule Codes

An index table contains the frequent itemsets of size greater than one and the corresponding itemset codes. These frequent itemsets are generated from different rulebases. Example 5.7, being a continuation of Example 5.5, illustrates the procedure of searching an association rule in the index table.

Example 5.7 Here we show how to construct an index table, Table 5.2.

Table 5.2 Index table for searching an association rule

Itemset	*AC*	*AB*	*AE*	*BC*	*DF*	*EF*	*FH*	*ABC*
Code	0	00	10	000	010	100	110	0000

The itemset code corresponding to AC is 0. The itemset code 0 corresponds to the set of association rules $\{A\rightarrow C, C\rightarrow A\}$. We would like to discuss the procedure for searching an association rule in the index table. Suppose we wish to search for the association rule corresponding to rule code 111. We apply binary search technique to find code 111. The binary search technique is based on the length of an itemset code. The search might end up at the fourth cell containing itemset code 000. Now, we apply sequential search towards the right side of the forth cell, since *value*(000) < *value*(111). We find that 111 is not present in the index table. But, the code 111 is positioned in-between 110 and 0000, since $|111| < |0000|$ and *value*(111) > *value*(110). We define *value* of a code ω as the numerical value of the code, i.e. $value(\omega) = (\omega)_{10}$. For example, *value* (010) = 2. Thus, the sequential search stops at the cell containing itemset code 110. In general, for a rule code ω, we get a consecutive pair of itemset codes ($code_1$, $code_2$) in the index table, such that $code_1 \leq \omega < code_2$. Then $code_1$ is the desired itemset code. Let *Y* be the desired itemset corresponding to the rule code ω. Then ω corresponds to an association rule generated by *Y*. Thus, the itemset code corresponding to the rule code 111 is 110. The frequent itemset corresponding to itemset code 110 is *FH*. Thus, the association rule corresponding rule code 111 is $H \rightarrow F$.

Initially, the binary search procedure finds an itemset code of desired length. Then it moves forward or backward sequentially till we get the desired itemset code. The algorithm for searching an itemset code is shown below.

Algorithm 5.1 Search for the itemset code corresponding to a rule code in the index table.

procedure *itemset-code-search* (ω, *T*, *i*, *j*)

Inputs:
ω: rule code (an ACP code)
T: index table
i: start index
j: end index

Outputs:
Index of the itemset code corresponding to ω

```
x = |ω|;
k = binary-search (x, T, i, j);
if (value(ω) ≥ value((T(k).code)) then
    q = forward-sequential-search (ω, T, k + 1, j);
else
    q = backward-sequential-search (ω, T, k – 1, i);
end if
return(q);
```

end procedure

The above algorithm is described as follows. The algorithm *itemset-code-search* (Adhikari and Rao 2007) searches index table *T* between the *i*-th and *j*-th cells and returns the index of the itemset code corresponding to the rule code ω. The procedure *binary-search* returns an integer *k* corresponding to rule code ω. If $value(\omega) \geq value((T(k).\text{code})$ then we search sequentially in *T* from index ($k + 1$) to *j*. Otherwise, we search sequentially in *T* from index ($k - 1$) down to *i*. Let there are *m* cells in the index table. Then binary search requires maximum $\lfloor \log_2(m) \rfloor + 1$ comparisons (Knuth 1973). The sequential search makes $O(1)$ comparison, since codes ω and $T(k)$.code are close and the search is performed only once. Therefore, algorithm *itemset-code-search* takes $O(\log(m))$ time.

Now, we need to find the association rule generated from the itemset corresponding to the itemset code returned by algorithm *itemset-code-search*. We consider a certain example to illustrate the procedure for identifying association rule for a given rule code. Let us consider the rule code 0100. Using the above technique, we determine that 0000 is the itemset code corresponding to rule code 0100. The itemset corresponding to the itemset code 0000 is *ABC*. The association rules generated from itemset *ABC* could be numbered as follows: 0-th association rule (i.e., $A \rightarrow BC$) has rule code 0000, 1-th association rule (i.e., $B \rightarrow AC$) has rule code 0001, and so on. Proceeding in this way, we find that the 4-th association rule (i.e., $AC \rightarrow B$) has rule code 0100.

We now find the association rule number corresponding to rule code ω. Let X be the itemset corresponding to rule code ω, and ν be the itemset code corresponding to X. Let $RB(X)$ be the set of all possible association rules generated by X. From Lemma 5.1, we have $|RB(X)| = 2^{|x|} - 2$, for $|X| \geq 2$. If $|\nu| = |\omega|$ then ω corresponds to $(\omega_{10} - \nu_{10})$-th association rule generated from X, where Y_{10} denote the decimal value corresponding to binary code Y. If $|\nu| < |\omega|$ then ω corresponds to $(2^{|\nu|} - \omega_{10} + \nu_{10})$-th association rule generated from X. In this case, $\nu = 0000$, $\omega = 0100$, and $X = ABC$. Thus ω corresponds to 4-th association rule generated from X.

Algorithm 5.2 Find itemset and association rule number corresponding to a rule code.

procedure *rule-generation* (*k*, *T*, *C*, *X*)

Input:
k: index
T: Index table
C: rule code (an ACP code)

Output:
Itemset *X* corresponding to *C*
Association rule number corresponding to *C*

let $X = T(k)$.itemset;
if ($|T(k)$.code$| = |C|$) **then**
return $(C_{10} - (T(k).\text{code})_{10})$;
else
return $(2^{|T(k).\text{code}|} - (C)_{10} + (T(k).\text{code})_{10})$;
end if
end procedure

We assume that the algorithm *itemset-code-search* returns *k* as the index of the itemset code corresponding to rule code *C*. Using index table *T* and *k*, the algorithm *rule-generation* returns the rule number and the itemset corresponding to rule code *C*. The itemset is returned through argument *X*, and rule number is returned by the procedure.

The ACP coding maintains an index table in main memory. We show an example to verify that the amount of space occupied by a rulebase (including the overhead of indexing) is significantly less than that of other techniques. We determine an overhead of maintaining index table in the following situation.

Example 5.8 We refer here to Example 5.7. We encounter 8 frequent itemsets in the index table. Let there are 10,000 items in the given databases. Therefore 14 bits are required to identify an item. Thus an amount of storage space would require for *AC* and *ABC* are equal to $2 \times 14 = 28$ bits, and $3 \times 14 = 42$ bits, respectively. The size of index file is the size of itemsets plus the size of itemset codes. In this case, the index table consumes $(28 \times 7 + 42 \times 1) + 21$ bits, i.e., we encounter 259 bits. The total space required (including the overhead of indexing) to represent *RB* is equal to (259 + 922) bits (as mentioned in Example 5.6) = 1,181 bits. Based on the running example, we compare the amounts of storage space required to represent *RB* using different rulebase representation techniques (Table 5.3).

Table 5.3 Amounts of storage space required for representing *RB* using different rulebase representation techniques

Technique for representing *RB*	TWO	SBV	ACP
Amount of space (bits)	7,104	1,770	1,181

We observe that the ACP coding consumes the least amount of space to represent *RB*. Let *OI*(*T*) be the overhead of maintaining index table using technique *T*. A technique without optimization (TWO) might not maintain index table separately. In this case, *OI*(TWO) = 0 bit. But, the ACP coding performs better than the TWO because ACP coding optimizes storage spaces for representing components of an association rule.

We describe here the data structures used in the algorithm for representing rulebases using ACP coding. A frequent itemset could be described by the following attributes: database identification, itemset and support. The frequent itemsets generated from RB_i are stored into array FIS_i, $i = 1, 2, \ldots, n$. We keep all the generated frequent itemsets into array *FIS*. Also, we have calculated *f*-value for every distinct frequent itemset *X* in *FIS* such that $|X| \geq 2$. The frequent itemsets and their *f*-values are stored into array *IS_Table*. We present below an algorithm (Adhikari and Rao 2007) for representing different rulebases using the ACP coding.

Algorithm 5.3 Represent rulebases using ACP coding.

procedure *ACP-coding* (*n*, *RB*)

Input:
n: number of databases
RB: union of rulebases
Output:
Coded association rules

```
let FIS = φ;
for i = 1 to n do
read RB_i from secondary storage;
generate FIS_i from RB_i;
```

```
05:  FIS = FIS ∪ FIS_i;
06:  end for
07:  let j = 1;
08:  let i = 1;
09:  while ( i ≤ |FIS| ) do
10:    if ( |FIS(i).itemset| ≥ 2 ) then
11:      compute f(X);
12:      IS_Table(j).itemset = X;
13:      IS_Table(j).f(X) = f(X);
14:      increase index j by 1;
15:      update index i for processing the next frequent itemset in FIS;
16:    end if
17:  end for
18:  sort itemsets in IS_Table using |X| as the primary key and f(X) as the secondary key;
19:  for i = 1 to |IS_Table| do
20:    C = ACP code of IS_Table (i).itemset;
21:    T(i) .itemset = IS_Table(i).itemset;
22:    T(i).code = C;
23:  end for
end procedure
```

In lines 1–6, we have generated frequent itemsets from different rulebases and are stored them into array *FIS*. We compute *f*-value for every frequent itemset *X* and store it into *IS_Table* (lines 7–17), for $|X| \geq 2$. At line 18, we sort frequent itemsets in *IS_Table* for the purpose of coding. Index table *T* is constructed by using lines 19–23.

Let *maximum* $\{|FIS_i|: 1 \leq i \leq n\}$ be p. Then the total number of itemsets is $O(n \times p)$. Therefore, lines 7–17 take $O(n \times p)$ time. Line 18 takes $O(n \times p \times \log(n \times p))$ time to sort $O(n \times p)$ itemsets. Lines 19–23 take $O(n \times p)$ time to construct the index table.

5.4.2 Storing Rulebases in Secondary Memory

An association rule could be stored in main memory using the following components: database identification, rule code, support, and confidence. Database identification, support and confidence could be stored using the method described in Section 5.3. Furthermore we need to maintain an index table in main memory to code/decode an association rule.

The rulebases corresponding to different databases could be stored in secondary memory using a bit sequential file *F*. The first line of *F* contains the number of databases. The second line of *F* contains the number of association rules in the first rulebase. The following lines of *F* contain the association rules in the first rulebase.

After keeping all the association rules in the first rulebase, we keep number of association rules in the second rulebase, and the association rules in the second rulebase thereafter. We illustrate the proposed file structure using the following example.

Example 5.9 Assume that there are 3 databases D_1, D_2, and D_3. Let the number of association rules extracted from these databases be 3, 4, and 2, respectively. The coded rulebases could be stored as follows:

<3><\n>
<3><\n>
<r11><s11><c11><\n>
<r12><s12><c12><\n>
<r13><s13><c13><\n>
<4><\n>
<r21><s21><c21><\n>
<r22><s22><c22><\n>
<r23><s23><c23><\n>
<r24><s24><c24><\n>
<2><\n>
<r31><s31><c31><\n>
<r32><s32><c32><\n>

"*n*" stands for the new line character. While storing an association rule in the secondary memory, if it contains a bit combination as that of "*n*", then we need to insert one more "*n*" after the occurrence of "*n*". We need not store the database identification along with an association rule, since the *i*-th set of association rules corresponds to the *i*-th database, $i = 1, 2, 3$. The terms r_{ij}, s_{ij}, and c_{ij} denote the rule code, support, and confidence of *j*-th association rule reported from *i*-th database, respectively, $j = 1, 2, \ldots, |RB_i|$, and $i = 1, 2, 3$.

Lemma 5.6 *Let M be the number of association rules with distinct antecedent-consequent pairs among N association rules reported from n databases, where* $2^{m-1} < M \leq 2^m$, *for an integer m. Suppose the support and confidence of an association rule are represented by fractions containing k digits after the decimal point. Assume that a frequent itemset X in FIS generates all possible association rules, for* $|X| \geq 2$. *Then the minimum amount of storage space required to represent RB in secondary memory is given as follows.*

$$SP^{\text{ACP coding}}_{\text{min, secondary}}(RB|\alpha, \beta) = 12 \times n + M \times (m-1) + N \times (2 \times \lceil k \times log_2 10 \rceil + 8) \\ -2 \times (2^{m-1} - m) + 12\,bits, \ if\ M < 2^m - 2;\ \ and$$

$$SP^{\text{ACP coding}}_{\text{min, secondary}}(RB|\alpha, \beta) = 12 \times n + M \times m + N \times (2 \times \lceil k \times log_2 10 \rceil + 8) \\ -2 \times (2^m - m - 1) + 12\,bits,\ \ otherwise.$$

Proof We do not need to store the database identification in the secondary storage, as the rulebases are stored sequentially one after another. A typical compiler represents

'\n' and an integer value using 1 byte and 4 bytes, respectively. The amount of memory required to represent the new line characters is equal to $P = 8 \times (N + n + 1)$ bits. The amount of memory required to store the number of databases and the number of association rules of each rulebase is equal to $Q = 4 \times (n + 1)$ bits. The amount of memory required to represent both the support and confidence of N rules is equal to $R = N \times 2\times \lceil k \times \log_2 10\rceil$ bits (as mentioned in Lemma 5.3). Let S be the minimum amount of memory required to represent the ACPs of M rules. Then, $S = M \times (m-1) - 2 \times (2^{m-1} - m)$ bits, if $M < 2^m - 2$, and $S = M \times m - 2 \times (2^m - m - 1)$ bits, otherwise (as mentioned in Lemma 5.4). Thus the minimum amount of storage space required to represent RB in the secondary memory is equal to $(P + Q + R + S)$ bits.

5.4.3 Space Efficiency of Our Approach

The effectiveness of a rulebase representation technique requires to be validated by its storage efficiency. There are many ways one could define the storage efficiency of a rulebase representation technique. We use the following definition to measure the storage efficiency of a rulebase representation technique.

Definition 5.1 Let RB_i be the rulebase corresponding to database D_i at a given pair (α, β), $i = 1, 2, \ldots, n$. Let RB be the union of rulebases corresponding to different databases. The space efficiency of technique T for representing RB is defined as follows:

$$\varepsilon\,(T, RB|\,\alpha, \beta, \psi) = \frac{SP_{\min}\,(RB|\alpha, \beta, \psi)}{SP^{T}\,(RB|\alpha, \beta, \psi)}, \text{ for } T \in \Gamma$$

The symbols and notation have been specified in Section 5.1.

We note that $0 < \varepsilon \leq 1$. We say that a rulebase representation technique is good if the value of ε is close to 1. We show that ACP coding stores rulebases at higher level of efficiency than that of any other representation technique.

Lemma 5.7 *Let RB_i be the set of association rules extracted from database D_i at a given pair (α, β), $i = 1, 2, \ldots, n$. Let RB be the union of rulebases corresponding to different databases. Also, let Γ be the set of all rulebase representation techniques. Then ϵ (ACP coding, $RB \mid \alpha, \beta, \psi) \geq \varepsilon(T, RB \mid \alpha, \beta, \psi)$, for $T \in \Gamma$.*

Proof We show that ACP coding stores RB using minimum storage space at a given pair (α, β). A local association rule has the following components: database identification, antecedent, consequent, support, and confidence. We classify the above components into the following three groups: {database identification}, {antecedent, consequent}, and {support, confidence}. Among these three groups, the item of group 1 is independent of the items of other groups. If there are n databases, we need a minimum of $\lceil \log_2 n\rceil$ bits to represent the item of group 1 (as shown in Lemma 5.2). Many association rules may have the same antecedent-consequent pair.

If an antecedent-consequent pair appears in many association rules, then it receives a shorter code. Therefore the antecedent-consequent pair of association rule having highest frequency is represented by a code of smallest size. ACP code starts from 0, and then follows the sequence 1, 00, 01, 10, 11, 000, 001, …. Therefore, no other technique would provide sizes of codes shorter than them. Therefore, the items of group 2 are expressed minimally using ACP codes. Again, the items of group 3 are related with the items of group 2. Suppose we keep p digits after the decimal point for representing an item of group 3. Then the representation an item of group 3 becomes independent of the one present for items of group 2. We need minimum $2 \times \lceil p \times \log_2 10 \rceil$ bits to represent support and confidence of an association rule (as mentioned in Lemma 5.3).

Thus, *minimum {representation of an association rule}* = *minimum {representation of items of group 1 + representation of items of group 2 + representation of items of group 3}* = *minimum {representation of items of group 1}* + *minimum {representation of items of group 2}* + *minimum{ representation of items of group 3}*.

Also, there will be an entry in the index table for the itemset corresponding to an association rule for the coding/decoding process.

Thus we have

minimum {representation of index table} = *minimum {representation of itemsets + representation of codes}*.

If there are p items then an itemset of size k could be represented by $k \times \lceil \log_2 (p) \rceil$ bits (as mentioned in Lemma 5.2). Also, ACP codes consume minimum space because of the way they have been designed.

Thus

minimum {representation of index table} = *minimum {representation of itemsets + minimum {representation of codes}*.

Therefore

minimum {representation of rulebases} = $\sum_r$ *{representation of association rule r using ACP coding}* + *representation of index table used in ACP coding.*

Hence the lemma follows.

Lemma 5.8 *Let RB_i be the set of association rules extracted from database D_i at a given pair (α, β), $i = 1, 2, \ldots, n$. Let RB be the union of rulebases corresponding to different databases. Then,* $SP_{\min}(RB \mid \alpha, \beta, \psi) = SP^{\text{ACPcoding}}_{\min}(RB \mid \alpha, \beta, \psi)$.

Proof From Lemma 5.7, we conclude that ACP coding represents rulebases using lesser amount of storage space than that of any other technique. Thus, $SP^{\text{ACPcoding}}_{\min}(RB \mid \alpha, \beta, \psi) \le SP^{T}_{\min}(RB \mid \alpha, \beta, \psi)$, for $T \in \Gamma$. We observe that a rulebase representation technique T might not represent rulebases at its minimum level because of the random nature of the set of transactions contained in a database. In other words, a frequent itemset may not generate all the association rules in a database. For example, the association rule $X \rightarrow Y$ may not get extracted from some of the given databases, even though the itemset $\{X, Y\}$ is frequent in the remaining databases. If the ACP coding represents *RB* using minimum storage space then it would be the minimum representation of *RB* at a given tuple (α, β, ψ).

There are many ways one could define the quality of synthesized patterns. We define the quality of synthesized patterns as follows.

Definition 5.2 Let RB_i be the rulebase extracted from database D_i at a given pair (α, β), $i = 1, 2, \ldots, n$. Let *RB* be the union of rulebases corresponding to different databases. We represent *RB* using a rulebase representation technique *T*. Let $\xi(RB \mid T, \alpha, \beta, \varphi, \psi)$ denote the collection of synthesized patterns over *RB* at a given tuple $(T, \alpha, \beta, \varphi, \psi)$. We define quality of $\xi(RB \mid T, \alpha, \beta, \varphi, \psi)$ as $\varepsilon(T, RB \mid \alpha, \beta, \varphi, \psi)$. The symbols and notation have been specified in Section 5.1.

Also, ε *(ACP coding, RB,* $\alpha, \beta) \geq \varepsilon$ (T, RB, α, β), for $T \in \Gamma$ (as mentioned in Lemma 5.7). Thus the quality of $\xi(RB \mid ACP\ coding, \alpha, \beta, \varphi, \psi) \geq$ quality of $\xi(RB \mid T, \alpha, \beta, \varphi, \psi)$, for $T \in \Gamma$.

5.5 Experiments

We have carried out several experiments to study the effectiveness of ACP coding. The following experiments are based on the transactional databases T10I4D100K (T_1) (Frequent itemset mining dataset repository 2004), and T40I10D100K (T_2) (Frequent itemset mining dataset repository 2004). These databases were generated using synthetic database generator from IBM Almaden Quest research group. We present some characteristics of these databases in Table 5.4.

Table 5.4 Database characteristics

Database	*N T*	*ALT*	*AFI*	*NI*
T_1	1,00,000	11.10	1,276.12	870
T_4	1,00,000	40.41	4,310.52	942

For the purpose of conducting the experiments, we divide each of these databases into 10 databases. We call these two sets of 10 databases as the input databases. The database T_i has been divided into 10 databases T_{ij} of size 10,000 transactions each, $j = 0, 1, 2, \ldots, 9$, and $i = 1, 4$. We present the characteristics of the input databases in Table 5.5.

The results of mining input databases are given in Table 5.6. The notations used in the above tables are explained as follows. *NT*, *ALT*, *AFI* and *NI* stand for number of transactions, average length of a transaction, average frequency of an item, and number of items in the data source, respectively. Some results are presented in Table 5.6.

In the above table, *NkIR* stands for the number of *k*-item association rules resulting from different databases, for $k \geq 2$. We present a comparison among different rulebase representation techniques, see Tables 5.7 and 5.8.

Table 5.5 Input database characteristics

Database	*ALT*	*AFI*	*NI*	Database	*ALT*	*AFI*	*NI*
T_{10}	11.06	127.66	866	T_{40}	40.57	431.56	940
T_{11}	11.13	128.41	867	T_{41}	40.58	432.19	939
T_{12}	11.07	127.65	867	T_{42}	40.63	431.79	941
T_{13}	11.12	128.44	866	T_{43}	40.63	431.74	941
T_{14}	11.14	128.75	865	T_{44}	40.66	432.56	940
T_{15}	11.14	128.63	866	T_{45}	40.51	430.46	941
T_{16}	11.11	128.56	864	T_{46}	40.74	433.44	940
T_{17}	11.10	128.45	864	T_{47}	40.62	431.71	941
T_{18}	11.08	128.56	862	T_{48}	40.53	431.15	940
T_{19}	11.081	128.11	865	T_{49}	40.58	432.16	939

Table 5.6 Results of data mining

Database	α	β	*N2IR*	*N3IR*	*NkIR* ($k > 3$)
$\bigcup_{i=1}^{10} T_{1i}$	0.01	0.2	136	29	0
$\bigcup_{i=1}^{10} T_{4i}$	0.05	0.2	262	0	0

Table 5.7 Different rulebase representation techniques-comparative analysis

Database	*SP(TWO)*	*SP(SBV)*	*OI*	*SP(ACP)*	*MSO*	*AC(SBV)*	*AC(ACP)*
$\bigcup_{i=1}^{10} T_{1i}$	48,448 bits	10,879 bits	619 bits	7,121 bits	7,051 bits	1.79640	1.17586
$\bigcup_{i=1}^{10} T_{4i}$	75,456 bits	16,768 bits	549 bits	10,681 bits	10,661 bits	1.77778	1.13242

Table 5.8 Comparison among different rulebase representation techniques

Database	ε(TWO)	ε(SBV)	ε(ACP)
$\bigcup_{i=1}^{10} T_{1i}$	0.14554	0.64813	0.99017
$\bigcup_{i=1}^{10} T_{4i}$	0.14129	0.63579	0.99813

In the above tables, we use the following abbreviations: *SP* stands for storage space (including overhead of indexing), *MSO* denotes the minimum storage space for representing rulebases including the overhead of indexing, and *AC*(*T*) stands for amount of compression (bits/byte) using technique *T*. In Fig. 5.1, we compare different rulebase representation techniques at different levels of minimum support.

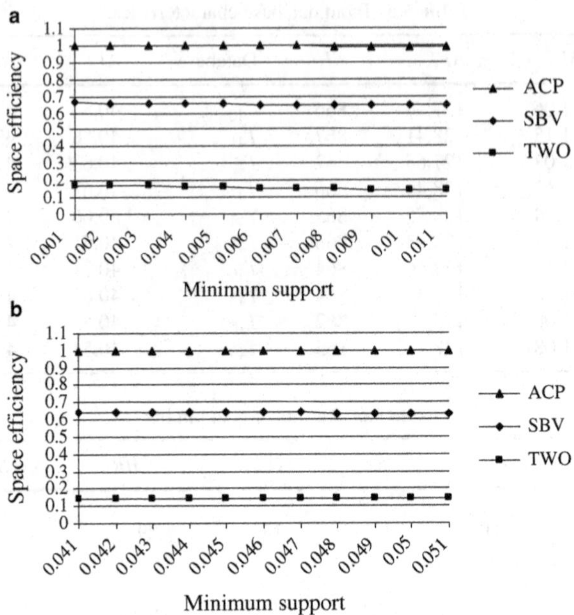

Fig. 5.1 Storage efficiency of different rulebase representation techniques. (**a**) For association rules extracted from T_{1i}, $i = 0, 1, \ldots, 9$. (**b**) For association rules extracted from T_{1i}, $i = 0, 1, \ldots, 9$

We have fixed the value β at 0.2 for all the experiments. The results show that the ACP coding stores rulebases most efficiently among different rulebase representation techniques. Also, we find that the SBV coding reduces the size of a rulebase considerably, but stores less efficiently than the ACP coding. This coding achieves maximum efficiency when the following two conditions are satisfied: (i) All the databases are of similar type and extract an identical set of association rules, and (ii) Each of the frequent itemsets of size greater than one generates all possible association rules.

Nelson (1996) studied data compression with the Burrows-Wheeler Transformation (BWT) (Burrows and Wheeler 1994). Experiments were carried out on 18 different files and average compression obtained by techniques using BWT and PKZIP are 2.41 and 2.64 bits/byte, respectively.

The results of Figs. 5.1(a) and 5.1(b) are carried out at 11 different pairs of values of pairs of (α, β). Using the ACP coding, we have obtained average compression 1.15014 bits/byte and 1.12190 bits/bytes for the experiments referring to Figs. 5.1(a) and 5.1(b), respectively.

5.6 Conclusions

An efficient storage representation of a set of pattern bases could contribute to the foundations of a multi-database mining system. Based on them, many applications of data mining of global nature could be developed in an efficient manner as reported through experimental results presented in this chapter. Similar technique could be employed to store frequent itemsets in different databases.

References

Adhikari A, Rao PR (2007) Enhancing quality of knowledge synthesized from multi-database mining. Pattern Recognition Letters 28(16): 2312–2324

Adhikari A, Rao PR (2008) Efficient clustering of databases induced by local patterns. Decision Support Systems 44(4): 925–943

Agrawal R, Imielinski T, Swami A (1993) Mining association rules between sets of items in large databases. In: Proceedings of ACM SIGMOD Conference, Washington, DC, pp. 207–216

Agrawal R, Srikant R (1994) Fast algorithms for mining association rules. In: Proceedings of International Conference on Very Large Data Bases, pp. 487–499

Ananthanarayana VS, Murty MN, Subramanian DK (2003) Tree structure for efficient data mining using rough sets. Pattern Recognition Letters 24(6): 851–862

Burrows M, Wheeler DJ (1994) A block-sorting lossless data compression algorithm. DEC, Digital Systems Research Center, Research Report 124

Bykowski A, Rigotti C (2003) A condensed representation to find frequent patterns for efficient mining. Information Systems 28(8): 949–977

Coenen F, Leng P, Ahmed S (2004) Data structure for association rule mining: T-trees and P-trees. IEEE Transactions on Knowledge and Data Engineering 16(6): 774–778

Fraleigh JB (1982) A First Course in Abstract Algebra. Third edition, Addision-Wesley, Reading, MA

Frequent Itemset Mining Dataset Repository (2004) http://fimi.cs.helsinki.fi/data

Han J, Pei J, Yiwen Y (2000) Mining frequent patterns without candidate generation. In: Proceedings of ACM SIGMOD Conference on Management of Data, Dallas, TX, pp. 1–12

Huffman DA (1952) A method for the construction of minimum redundancy codes. In: Proceedings of the IRE 40(9), pp. 1098–1101

Jeudy B, Boulicaut JF (2002) Using condensed representations for interactive association rule mining. In: Proceedings of PKDD, LNAI 2431, Helsinki, FIN, pp. 225–236

Knuth DE (1973) The Art of Computer Programming. Volume 3, Addision-Wesley, Reading, MA

Kum H-C, Chang HC, Wang W (2006) Sequential pattern mining in multi-databases via multiple alignment. Data Mining and Knowledge Discovery 12(2–3): 151–180

Nelson MR (1996) Data compression with the Burrows-Wheeler transformation. Dr. Dobb's Journal (September): 46–50

Pasquier N, Taouil R, Bastide Y, Stumme G, Lakhal L (2005) Generating a condensed representation for association rules. Journal of Intelligent Information Systems 24(1): 29–60

Savasere A, Omiecinski E, Navathe S (1995) An efficient algorithm for mining association rules in large databases. In: Proceedings of the 21st International Conference on Very Large Data Bases, pp. 432–443

Sayood K (2000) Introduction to Data Compression. Morgan Kaufmann, San Francisco

Shenoy P, Haritsa JR, Sudarshan S, Bhalotia G, Bawa M, Shah D (2000) Turbo-charging vertical mining of large databases. In: Proceedings of ACM SIGMOD Conference on Management of Data, Dallas, TX, pp. 22–33

Su K, Huang H, Wu X, S. Zhang S (2006) A logical framework for identifying quality knowledge from different data sources. Decision Support Systems 42(3): 1673–1683

Wu X, Zhang S (2003) Synthesizing high-frequency rules from different data sources. IEEE Transactions on Knowledge and Data Engineering 14(2): 353–367

Wu X, Zhang C, Zhang S (2005) Database classification for multi-database mining. Information Systems 30(1): 71–88

Zaki MJ, Ogihara M (1998) Theoretical foundations of association rules. In: Proceedings of the DMKD Workshop on Research Issues in Data Mining and Knowledge Discovery, New York, pp. 7:1–7:8

Zhang S (2002) Knowledge discovery in multi-databases by analyzing local instances, Ph D thesis, Deakin University

Zhang C, Liu M, Nie W, Zhang S (2004) Identifying global exceptional patterns in multi-database mining. IEEE Computational Intelligence Bulletin 3(1): 19–24

Zhang S, Zaki MJ (2002) Mining Multiple Data Sources: Local Pattern Analysis. Data Mining and Knowledge Discovery, Springer, New York, pp. 121–125

Chapter 6
Efficient Clustering of Databases Induced by Local Patterns

In view of answering queries provided in multiple large databases, it might be required to mine relevant databases *en block*. In this chapter, we present an efficient solution to clustering multiple large databases. We present two measures of similarity between a pair of databases and study their main properties. In the sequel, we design an algorithm for clustering multiple databases based on an introduced similarity measure. Also, we present a coding, referred to as IS coding, to represent itemsets space efficiently. The coding of this nature enables more frequent itemsets to participate in the determination of the similarity between two databases. Thus the invoked clustering process becomes more accurate. We also show that the IS coding attains maximum efficiency in most of the cases of the mining processes. The clustering algorithm becomes improved (in terms of its time complexity) when contrasted with the existing clustering algorithms. The efficiency of the clustering process has been improved using several strategies that is by reducing execution time of the clustering algorithm, using more suitable similarity measure, and storing frequent itemsets space efficiently.

6.1 Introduction

Effective data analysis using a traditional data mining technique on multi-gigabyte repositories has proven difficult. A quick discovery of approximate knowledge from large databases would be adequate for many decision support applications.

As before, let us consider a company that deals with multiple large databases. The company might need to carry out association analysis involving non-profit making items (products). The ultimate objective is to identify the items that neither make much profit nor help promoting other products. An association analysis involving non-profit making items might identify such items. The company could then stop dealing with them. The analysis of this nature might require identifying similar databases. Let us note that two databases are deemed similar if they contain many similar transactions. Again, two transactions are similar if they have many common items. We observe later that two databases containing many common items are not necessarily very similar. First, let us define a few concepts used frequently in this chapter.

A. Adhikari et al., *Developing Multi-database Mining Applications*, Advanced Information and Knowledge Processing, DOI 10.1007/978-1-84996-044-1_6,

Let $I(D)$ be the set of items in database D. An *itemset* is a set of items in a database. An itemset X in D is associated with a statistical measure called support (Agrawal et al. 1993), denoted by $supp(X, D)$, for $X \subseteq I(D)$. *Support* of an itemset X in D is the fraction of transactions in D containing X. The importance of an itemset could be judged by quantifying its support. X is called a *frequent itemset* (*FIS*) in D if $supp(X, D) \geq \alpha$, where α is the user-defined *minimum support*. A frequent itemset possesses higher support. Thus the collection of frequent itemsets determines major characteristics of a database. One could define similarity between a pair databases in terms of their frequent itemsets. We may observe that two databases are similar if they have many common frequent itemsets.

Based on the similarity between two databases, one could cluster branch databases. Once the clustering process has been completed, one could mine all the databases in a class together to make an approximate association analysis involving frequent items. An approximate association analysis could be performed using the frequent itemsets in the union of all the databases in a class. In this manner, clustering of databases helps reducing data for analyzing the items. In what follows, we study the problem of clustering transactional databases using the local frequent itemsets.

For clustering transactional databases, Wu et al. (2005b) have proposed two similarity measures, denoted as sim_1, and sim_2. Let $D = \{D_1, D_2, \ldots, D_n\}$, where D_i is the database corresponding to the i-th branch of a multi-branch company, $i = 1, 2, \ldots, n$. sim_1 is based on the items present in the databases, and becomes defined as follows:

$$sim_1(D_1, D_2) = |I(D_1) \cap I(D_2)| \;/\; |I(D_1) \cup I(D_2)|$$

Let S_i be the set of association rules present in D_i, $i = 1, 2, \ldots, n$. The measure sim_2 is based on the items generated from S_i, $i = 1, 2, \ldots, n$. Let $I(S_i)$ be the set of items generated from S_i, $i = 1, 2, \ldots, n$. The similarity measure sim_2 is expressed in the form:

$$sim_2(D_1, D_2) = |I(S_1) \cap I(S_2)| \;/\; |I(S_1) \cup I(S_2)|$$

$I(S_i) \subseteq I(D_i)$, for $i = 1, 2, \ldots, n$. sim_1 estimates similarity between two databases more correctly than sim_2, since the number of items which participate in determining the value of the similarity between two databases under sim_1 is higher than that of sim_2. A database may not extract any association rule for given values of (α, β), where β is the user-defined *minimum confidence level*. In such situations, the accuracy of sim_2 is low. In the following example, we discuss a situation where the accuracy of sim_1 and sim_2 are low.

Example 6.1 Consider a multi-branch company that possesses following three databases:

$DB_1 = \{\ \{a, b, c, e\}, \{a, b, d, f\}, \{b, c, g\}, \{b, d, g\}\ \}$
$DB_2 = \{\ \{a, g\}, \{b, e\}, \{c, f\}, \{d, g\}\ \}$ and
$DB_3 = \{\ \{a, b, c\}, \{a, b, d\}, \{b, c\}, \{b, d, g\}\ \}$

Here, $I(DB_1) = \{a, b, c, d, e, f, g\}$, $I(DB_2) = \{a, b, c, d, e, f, g\}$, $I(DB_3) = \{$a, b, c, d, g$\}$. Thus, $sim_1(DB_1, DB_2) = 1.0$ (maximum), and $sim_1(DB_1, DB_3) = 0.71429$. Ground realities are as follows: (i) The similarity between DB_1 and DB_2 is low, since they contain dissimilar transactions. (ii) The similarity between DB_1 and DB_3 is higher than the similarity observed between DB_1 and DB_2, since DB_1 and DB_3 contain similar transactions. Hence the similarity measures sim_1 produces low accuracy in finding the similarity between two databases. There are no frequent itemsets in DB_2, if $\alpha > 0.25$. Thus, $I(S_2) = \phi$, if $\alpha > 0.25$. Hence, the accuracy of sim_2 is low in finding the similarity between DB_1 and DB_2 if $\alpha > 0.25$.

We have observed that the similarity measures based on items in databases might not be appropriate in finding similarity between two databases. A more suitable similarity measure could be designed based on frequent itemsets present in both the databases. The frequent itemsets in two databases could find better the similarity among transactions in two databases. Thus, frequent itemsets in two databases could find similarity between two databases correctly.

Wu et al. (2005a) have proposed a solution of inverse frequent itemset mining. They argued that one could efficiently generate a synthetic market basket database from the frequent itemsets and their supports. Thus, the similarity between two databases could be estimated correctly by involving supports of the frequent itemsets. We propose two measures of similarity based on the frequent itemsets and their supports. A new algorithm for clustering databases is designed based on a proposed measure of similarity.

The existing industry practice is to refresh a data warehouse on a periodic basis. Let λ be the frequency of this process of data warehouse refreshing. In this situation, an incremental mining algorithm (Lee et al. 2001) could be used to obtain updated supports of the existing frequent itemsets in a database. But, there could be addition or, deletion of frequent itemsets over time. We need to mine the databases individually and again this is being done in a periodic manner. Let Λ be the periodicity of data warehouse mining. The values of λ and Λ could be chosen in such way that $\Lambda > \lambda$. Based on the updated local frequent itemsets, one could cluster the databases afresh.

Another alternative for taming multi-gigabyte data could be sampling. Let us note that a commonly used technique for approximate query answering is sampling (Babcock et al. 2003). If an itemset is frequent in a large database then it is likely that this itemset is frequent in a sample database. Thus, one could analyze approximately a database by analyzing the frequent itemsets present in a sample database.

The chapter is organized so that it reflects the main objectives identified above. We formulate the problem in Section 6.2. In Section 6.3, we discuss some related work. In Section 6.4, we show how to cluster all the branch databases. The experimental results are presented in Section 6.5.

6.2 Problem Statement

Let there are n branch databases. Also, let $FIS(D_i, \alpha)$ be the set of frequent itemsets corresponding to database D_i at a given value of α, $i = 1, 2, \ldots, n$. The problem is stated succinctly as follows:

Find the best non-trivial partition (if it exists) of $\{D_1, D_2, \ldots, D_n\}$ *using* $FIS(D_i, \alpha)$, $i = 1, 2, \ldots, n$.

A partition (Liu 1985) is a specific type of clustering. A formal definition of a non-trivial partition will be given in Section 6.4.

6.2.1 Related Work

Jain et al. (1999) have presented an overview of clustering methods from a statistical pattern recognition perspective, with a goal of providing a useful advice and references to fundamental concepts accessible to the broad community of clustering practitioners. A traditional clustering technique (Zhang et al. 1997) is based on *metric* attributes. A *metric* attribute is one whose values can be represented by explicit coordinates in a Euclidean space. Thus a traditional clustering technique might not work in this case, since we are interested in clustering databases. Ali et al. (1997) have proposed a partial classification technique using association rules. The clustering of databases using local association rules might not be a good idea. The number of frequent itemsets obtained from a set of association rules might be much less than the number of frequent itemsets extracted using the apriori algorithm (Agrawal et al. 1993). In this way, the efficiency of the clustering process could be low. Liu et al. (2001) have proposed a multi-database mining technique that searches only the relevant databases. Identifying relevant databases is based on selecting the relevant tables (relations) that contain specific, reliable and statistically significant information pertaining to the query. Our study involves clustering transactional databases. Yin and Han (2005) have proposed a new strategy for relational heterogeneous database classification. This strategy might not be suitable for clustering transactional databases. Yin et at. (2006) have proposed two scalable methods for multi-relational classification: CrossMine-Rule, a rule-based method and CrossMine-Tree, a decision-tree-based method. Bandyopadhyay et al. (2006) have proposed a technique for clustering homogeneously distributed data in a peer-to-peer environment like sensor networks. It is based on the idea of the K-Means clustering. It works in a localized asynchronous manner by realizing a communication with the neighboring nodes.

In the context of similarity measures, Tan et al. (2002) have presented an overview of twenty one interestingness measures available in statistics, machine learning, and data mining literature. Support and confidence measures (Agrawal et al. 1993) are used to identify frequently occurring association rules between two sets of items in large databases. Our first measure, $simi_1$, is similar to the Jaccard measure (Tan et al. 2002). Measures such as support, interest (Tan et al. 2002), cosine (Tan et al. 2002) are expressed as a ratio of two quantities. Their numerators represent a kind of closeness between two objects. But their denominators are not appropriate to make these ratios as measures of association. As a result they do not serve as sound measures of similarity.

Zhang et al. (2003) designed a local pattern analysis for mining multiple databases. Zhang (2002) studied various strategies for mining multiple databases.

For utilizing low-cost information and knowledge on the internet, Su et al. (2006) have proposed a logical framework for identifying knowledge of sound quality coming from different data sources. It helps working towards the development of a generally acceptable ontology.

Privacy concerns over sensitive data have become important in knowledge discovery. Usually, data owners have different levels of concerns over different data attributes, which adds complexity to data privacy. Moreover, collusion among malicious adversaries poses a severe threat to data security. Yang and Huang (2008) have proposed an efficient clustering method for distributed multi-party data sets using the orthogonal transformation and perturbation techniques. It allows data owners to set up different levels of privacy for different attributes.

In many large e-commerce organizations, multiple data sources are often used to describe the same customers, thus it is important to consolidate data of multiple sources for intelligent business decision making. Ling and Yang (2006) have proposed a method that predicts the classification of data from multiple sources without class labels in each source.

6.3 Clustering Databases

The approach of finding the best partition of a set of databases can be explained through a sequence of the following steps:

(i) Find $FIS(D_i, \alpha)$, for $i = 1, 2, \ldots, n$.
(ii) Determine the similarity between each pair of databases using the proposed measure of similarity $simi_2$.
(iii) Check for the existence of partitions at the required similarity levels (as mentioned in Theorem 6.5).
(iv) Calculate the goodness values for all the non-trivial partitions.
(v) Report the non-trivial partition for which the goodness value attains its maximum.

The steps (i)–(v) will be followed and explained with the help of a running example. We start with an example of a multi-branch company that has multiple databases.

Example 6.2 A multi-branch company has seven branches. The branch databases are given below.

$$D_1 = \{(a, b, c), (a, c), (a, c, d)\}$$
$$D_2 = \{(a, c), (a, b), (a, c, e)\}$$
$$D_3 = \{(a, e), (a, c, e), (a, b, c)\}$$
$$D_4 = \{(f, d), (f, d, h), (e, f, d), (e, f, h)\}$$
$$D_5 = \{(g, h, i), (i, j), (h, i), (i, j, g)\}$$
$$D_6 = \{(g, h, i), (i, j, h), (i, j)\}$$
$$D_7 = \{ (a, b), (g, h), (h, i), (h, i, j) \}$$

The sets of frequent itemsets are shown below:

$FIS(D_1, 0.35) = \{ (a, 1.0), (c, 1.0), (ac, 1.0) \}$
$FIS(D_2, 0.35) = \{ (a, 1.0), (c, 0.67), (ac, 0.67) \}$
$FIS(D_3, 0.35) = \{ (a, 1.0), (c, 0.67), (e, 0.67), (ac, 0.67), (ae, 0.67) \}$
$FIS(D_4, 0.35) = \{ (d, 0.75), (e, 0.5), (f, 1.0), (h, 0.5), (df, 0.75), (ef, 0.5), (fh, 0.5) \}$
$FIS(D_5, 0.35) = \{ (g, 0.5), (h, 0.5), (i, 1.0)\}, (j, 0.5), , (gi, 0.5), (hi, 0.5), (ij, 0.5) \}$
$FIS(D_6, 0.35) = \{ (i, 1.0), (j, 0.67), (h, 0.67), (hi, 0.67), (ij, 0.67) \}$
$FIS(D_7, 0.35) = \{ (h, 0.75), (i, 0.5), (hi, 0.5) \}$.

Based on the sets of frequent itemsets in a pair of databases, one could define many measures of similarity between them. The two measures of similarity between a pair of databases are suitable for dealing with the problem at hand. The first measure $simi_1$ (Adhikari and Rao 2008) is defined as follows:

Definition 6.1 The measure of similarity $simi_1$ between databases D_1 and D_2 is defined as the following ratio:

$$simi_1(D_1,D_2,\alpha) = \frac{|FIS(D_1,\alpha) \cap FIS(D_2,\alpha)|}{|FIS(D_1,\alpha) \cup FIS(D_2,\alpha)|},$$

where the symbols $\cup$ and $\cap$ stand for the intersection and union operations used in set theory, respectively.

The similarity measure $simi_1$ is the ratio of the number frequent itemsets common to D_1 and D_2, and the total number of distinct frequent itemsets in D_1 and D_2. Frequent itemsets are the dominant patterns that determine major characteristics of a database. There are many implementations of mining frequent itemsets in a database (FIMI 2004). Let X and Y be two frequent itemsets in database DB. The itemset X is more dominant than the itemset Y in DB if $supp(X, DB) > supp(Y, DB)$. Therefore the characteristics of DB are revealed more by the pair $(X, supp(X, DB))$ than that of $(Y, supp(Y, DB))$. In other words, a sound measure of similarity between two databases is a function of the supports of the frequent itemsets in the databases.

The second measure of similarity $simi_2$ (Adhikari and Rao 2008) comes in the form:

Definition 6.2 The measure of similarity $simi_2$ between databases D_1 and D_2 is defined as follows:

$$simi_2(D_1, D_2, \alpha) = \frac{\sum_{X \in \{FIS(D_1, \alpha) \cap FIS(D_2, \alpha)\}} minimum\{supp(X, D_1), supp(X, D_2)\}}{\sum_{X \in \{FIS(D_1, \alpha) \cup FIS(D_2, \alpha)\}} maximum\{ supp(X, D_1), supp(X, D_2)\}},$$

Here we assume that $supp(X, D_i) = 0$, if $X \notin FIS(D_i, \alpha)$, for $i = 1, 2$.

With reference to Example 6.1, the frequent itemsets in different databases are given as follows:

$FIS(DB_1, 0.3) = \{a(0.5), b(1.0), c(0.5), d(0.5), g(0.5), ab(0.5), bc(0.5), bd(0.5)\}$
$FIS(DB_2, 0.3) = \{g(0.5)\}$
$FIS(DB_3, 0.3) = \{a(0.5), b(1.0), c(0.5), d(0.5), ab(0.5), bc(0.5), bd(0.5)\}$

We obtain $simi_1(DB_1, DB_2, 0.3) = 0.125$, $simi_1(DB_1, DB_3, 0.3) = 0.875$, $simi_2(DB_1, DB_2, 0.3) = 0.111$, and $simi_2(DB_1, DB_3, 0.3) = 0.889$. Thus, the proposed measures $simi_1$ and $simi_2$ are more suitable than the existing measures mentioned in Example 6.1.

Theorem 6.1 justifies the fact that $simi_2$ is more appropriate measure than $simi_1$.

Theorem 6.1 *The similarity measure* $simi_2$ *exhibits higher discriminatory power than that of the similarity measure* $simi_1$.

Proof The support of a frequent itemset could be considered as its weight in the database. We attach weight 1.0 to itemset X in database D_i, under the similarity measure $simi_1$, if $X \in FIS(D_i, \alpha)$, $i = 1, 2$. We attach an weight $supp(X, D_i)$ to the itemset X in database D_i, under the similarity measure $simi_2$, if $X \in FIS(D_i, \alpha)$, $i = 1, 2$. The similarity measures sim_1 and sim_2 are defined as a ratio of two quantities. If $X \in FIS(D_i, \alpha)$, and $X \in FIS(D_j, \alpha)$, then it is more justifiable to add *minimum* $\{ supp(X, D_i), supp(X, D_j) \}$ (instead of 1.0) in the numerator and *maximum* $\{ supp(X, D_i), supp(X, D_j) \}$ (instead of 1.0) in the denominator for the itemset X, $i, j \in \{1, 2\}$. If $X \in FIS(D_i, \alpha)$, and $X \notin FIS(D_j, \alpha)$, then it is more justifiable to add 0 in the numerator and $supp(X, D_i)$ (instead of 1.0) in the denominator for itemset X, $i, j \in \{1, 2\}$. Hence, the theorem has been proved.

In Example 6.3, we verify that $simi_2$ is more appropriate measure than $simi_1$.

Example 6.3 With reference to Example 6.2, $supp(\{a\}, D_1) = supp(\{c\}, D_1) = supp(\{a, c\}, D_1) = 1.0$, $supp(\{a\}, D_2) = 1.0$, and $supp(\{c\}, D_2) = supp(\{a, c\}, D_2) = 0.67$. $simi_2(D_1, D_2, 0.35) = 0.78$, and $simi_1(D_1, D_2, 0.35) = 1.0$. We observe that the databases D_1 and D_2 are highly similar, but they are not the same. Thus, the similarity computed by $simi_2$ is more suitable.

We highlight some interesting properties of $simi_1$ and $simi_2$ by presenting Theorems 6.2, 6.3 and 6.4.

Theorem 6.2 *The similarity measure* $simi_k$ *satisfies the following properties* ($k = 1, 2$), (i) $0 \le simi_k(D_i, D_j, \alpha) \le 1$, (ii) $simi_k(D_i, D_j, \alpha) = simi_k(D_j, D_i, \alpha)$, (iii) $simi_k(D_i, D_i, \alpha) = 1$, *for* $i, j = 1, 2, \ldots, n$.

Proof The properties follow from the definition of $simi_k$, ($k = 1, 2$).

Now we express the distance between two databases in term of their similarity.

Definition 6.3 The distance measure $dist_k$ between two databases D_1 and D_2 based on the similarity measure $simi_k$ is defined as $dist_k(D_1, D_2, \alpha) = 1 - simi_k(D_1, D_2, \alpha)$, ($k = 1, 2$).

A "meaningful" distance satisfies the metric properties (Barte 1976). The higher the distance between two databases, the lower is the similarity between them. For the purpose of concise presentation, we use the notation I_i in place of $FIS(D_i, \alpha)$ used so far in Theorems 6.3 and 6.4, for $i = 1, 2$.

Theorem 6.3 *$dist_1$ is a metric over* [0, 1].

Proof We show that $dist_1$ satisfies the triangular inequality. Other properties of a metric follow from Theorem 6.2.

$$dist_1(D_1,D_2,\alpha) = 1 - \frac{|I_1 \cap I_2|}{|I_1 \cup I_2|} \geq \frac{|I_1 - I_2| + |I_2 - I_1|}{|I_1 \cup I_2 \cup I_3|} \tag{6.1}$$

Thus,

$$dist_1(D_1,D_2,)+dist_1(D_2,D_3,\alpha) \geq \frac{|I_1-I_2| + |I_2-I_1| + |I_2-I_3| + |I_3-I_2|}{|I_1 \cup I_2 \cup I_3|} \tag{6.2}$$

$$= \frac{|I_1 \cup I_2 \cup I_3| - |I_1 \cap I_2 \cap I_3| + |I_1 \cap I_3| + |I_2| - |I_1 \cap I_2| - |I_2 \cap I_3|}{|I_1 \cup I_2 \cup I_3|} \tag{6.3}$$

$$= 1 - \frac{|I_1 \cap I_2 \cap I_3| - |I_1 \cap I_3| - |I_2| + |I_1 \cap I_2| + |I_2 \cap I_3|}{|I_1 \cup I_2 \cup I_3|} \tag{6.4}$$

$$= 1 - \frac{\{|I_1 \cap I_2 \cap I_3| + |I_1 \cap I_2| + |I_2 \cap I_3|\} - \{|I_1 \cap I_3| + |I_2|\}}{|I_1 \cup I_2 \cup I_3|} \tag{6.5}$$

Let the number of elements in the shaded regions of Figs. 6.1(c) and 6.1(d) be N_1 and N_2, respectively. Then the expression (6.5) becomes

$$1 - \frac{N_1 - N_2}{|I_1 \cup I_2 \cup I_3|} \geq \begin{cases} 1 - \frac{N_1 - N_2}{|I_1 \cup I_2 \cup I_3|}, \text{ if } N_1 \geq N_2 & \text{(case 1)} \\ 1 - \frac{|I_1 \cap I_3|}{|I_1 \cup I_2 \cup I_3|}, \text{ if } N_1 < N_2 & \text{(case 2)} \end{cases} \tag{6.6}$$

In case 1, the expression remains the same. In case 2, a positive quantity $|I_1 \cap I_3|$ has been put in place of a negative quantity $N_1 - N_2$. Thus, the expression (6.6) reads as

$$\geq \begin{cases} 1 - \frac{N_1 - N_2}{|I_1 \cup I_3|}, \text{ if } N_1 \geq N_2 \\ 1 - \frac{|I_1 \cap I_3|}{|I_1 \cup I_3|}, \text{ if } N_1 < N_2 \end{cases} \geq \begin{cases} 1 - \frac{N_1}{|I_1 \cap I_3|}, \text{ if } N_1 \geq N_2 \\ 1 - \frac{|I_1 \cap I_3|}{|I_1 \cup I_3|}, \text{ if } N_1 < N_2 \end{cases} \tag{6.7}$$

$$\geq \begin{cases} 1 - \frac{|I_1 \cap I_3|}{|I_1 \cup I_3|}, \text{ if } N_1 \geq N_2 \\ 1 - \frac{|I_1 \cap I_3|}{|I_1 \cup I_3|}, \text{ if } N_1 < N_2 \end{cases}, \text{ where, } N_1 = |I_1 \cap I_2 \cap I_3| \leq |I_1 \cap I_3| \tag{6.8}$$

Therefore, irrespective of the relationship between N_1 and N_2, $dist_1(D_1, D_2, \alpha)$ + $dist_1(D_2, D_3, \alpha) \geq dist_1(D_1, D_3, \alpha)$. Thus, $dist_1$ satisfies the triangular inequality.

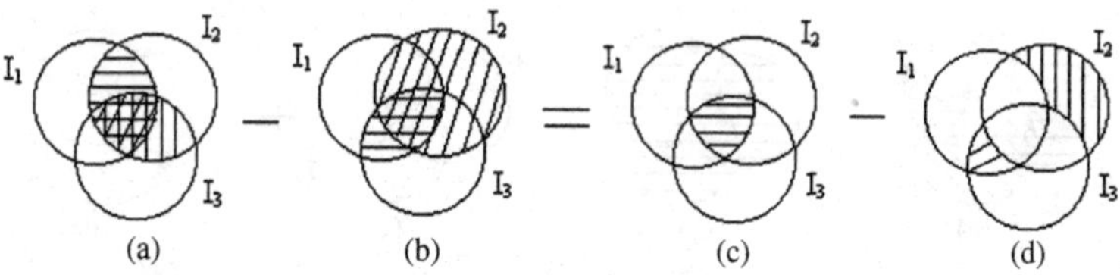

Fig. 6.1 Simplification of the expression (6.5) using Venn diagram

We also show that $dist_2$ satisfies the metric properties.

Theorem 6.4 *$dist_2$ is a metric over* [0, 1].

Proof We show that $dist_2$ satisfies the triangular inequality. The remaining properties of a metric follow from Theorem 6.2.

$$\begin{aligned} dist_2(D_1, D_2, \alpha) &= 1 - \frac{\sum_{x \in I_1 \cap I_2} minimum\{supp(x, D_1), supp(x, D_2)\}}{\sum_{x \in I_1 \cup I_2} maximum\{supp(x, D_1), supp(x, D_2)\}} \\ &= 1 - \frac{\sum_{x \in I_1 \cap I_2} min_{12}(x)}{\sum_{x \in I_1 \cup I_2} max_{12}(x)} \end{aligned} \tag{6.9}$$

where, $max_{ij}(x) = maximum\{\ supp(x, D_i), supp(x, D_j)\ \}$, and $min_{ij}(x) = minimum\ \{supp(x, D_i), supp(x, D_j)\ \}$, for $i \neq j$. Also, let $max_{123}(x) = maximum\{\ supp(x, D_1), supp(x, D_2), supp(x, D_3)\ \}$, and $min_{123}(x) = minimum\{\ supp(x, D_1), supp(x, D_2), supp(x, D_3)\ \}$.

Thus, $dist_2(D_1, D_2, \alpha) + dist_2(D_2, D_3, \alpha)$

$$= \frac{\sum_{x \in I_1 \cup I_2} max_{12}(x) - \sum_{x \in I_1 \cap I_2} min_{12}(x)}{\sum_{x \in I_1 \cup I_2} max_{12}(x)} + \frac{\sum_{x \in I_2 \cup I_3} max_{23}(x) - \sum_{x \in I_2 \cap I_3} min_{23}(x)}{\sum_{x \in I_2 \cup I_3} max_{23}(x)} \tag{6.10}$$

$$\geq \frac{\sum_{x \in I_1 - I_2} max_{12}(x) + \sum_{x \in I_2 - I_1} max_{12}(x)}{\sum_{x \in I_1 \cup I_2} max_{12}(x)} + \frac{\sum_{x \in I_2 - I_3} max_{23}(x) + \sum_{x \in I_3 - I_2} max_{23}(x)}{\sum_{x \in I_2 \cup I_3} max_{23}(x)} \tag{6.11}$$

$$\geq \frac{\sum_{x \in I_1 - I_2} max_{12}(x) + \sum_{x \in I_2 - I_1} max_{12}(x) + \sum_{x \in I_2 - I_3} max_{23}(x) + \sum_{x \in I_3 - I_2} max_{23}(x)}{\sum_{x \in I_1 \cup I_2 \cup I_3} max_{123}(x)} \tag{6.12}$$

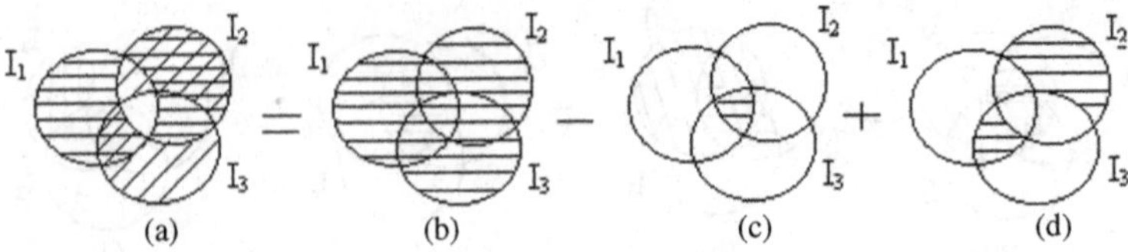

Fig. 6.2 Simplification of the expression (6.12) using Venn diagram

Using the simplification visualized graphically in Fig. 6.2, the expression (6.12) becomes

$$\frac{\sum_{x \in I_1 \cup I_2 \cup I_3} max_{123}(x) - N_1 + N_2}{\sum_{x \in I_1 \cup I_2 \cup I_3} max_{123}(x)} \tag{6.13}$$

where N_1 and N_2 are the value of $\sum_x max_{123}(x)$ over the shaded regions of Figs. 6.2(c) and 6.2(d), respectively. The expression (6.13) is equal to

$$1 - \frac{N_1 - N_2}{\sum_{x \in I_1 \cup I_2 \cup I_3} max_{123}(x)} \geq \begin{cases} 1 - \frac{N_1}{\sum_{x \in I_1 \cup I_2 \cup I_3} max_{123}(x)}, & \text{if } N_1 \geq N_2 \\ 1 - \frac{N_1 - N_2}{\sum_{x \in I_1 \cup I_2 \cup I_3} max_{123}(x)}, & \text{if } N_1 < N_2 \end{cases}$$

$$\geq \begin{cases} 1 - \frac{\sum_{x \in I_1 \cap I_3} max_{13}(x)}{\sum_{x \in I_1 \cup I_2 \cup I_3} max_{123}(x)}, & \text{if } N_1 \geq N_2 \\ 1 - \frac{\sum_{x \in I_1 \cap I_3} max_{13}(x)}{\sum_{x \in I_1 \cup I_2 \cup I_3} max_{123}(x)}, & \text{if } N_1 < N_2 \end{cases} \tag{6.14}$$

Therefore, irrespective of the relationship between N_1 and N_2, $dist_2(D_1, D_2, \alpha)$ + $dist_2(D_2, D_3, \alpha) \geq dist_2(D_1, D_3, \alpha)$. Thus, $dist_2$ satisfies the triangular inequality.

Given a set of databases, the similarity between a collection of pairs of databases could be expressed by a square matrix, called *database similarity matrix* (DSM). We define DSM of a set of databases as follows:

Definition 6.4 Let $D = \{D_1, D_2, \ldots, D_n\}$ be the set of all databases. The database similarity matrix DSM_k of D expressed by the measure of similarity $simi_k$, is a symmetric square matrix of size n by n, whose (i, j)-th element $DSM_k^{i,j}(D, \alpha) = simi_k(D_i,$ $D_j, \alpha)$; for $D_i, D_j \in D$, and $i, j = 1, 2, \ldots, n$, $(k = 1, 2)$.

For n databases, there are ${}^nC_2$ pairs of databases. For each pair of databases, we determine the calculations of similarity between them. If the similarity is high then the databases may be placed in the same class. We define this class as follows:

Definition 6.5 Let $D = \{D_1, D_2, \ldots, D_n\}$. A class $class_k^\delta$ formed at the level of similarity δ under the measure of similarity $simi_k$, is defined as $class_k^\delta(D,\alpha) = \begin{cases} P{:}P \subseteq D, |P| \geq 2, \text{and } simi_k(A,B,\alpha) \geq \delta, \text{for } A,B \in P \\ P{:}P \subseteq D, |P| = 1 \end{cases}, (k = 1, 2)$.

A *DSM* could be viewed as a complete weighted graph. Each database forms a vertex of the graph. A weight of the edge is the similarity between the pair of the corresponding databases. During the process of clustering, we assume that the databases $D_1, D_2, \ldots, D_r$ have been included in some classes, and the remaining databases are yet to be clustered. Then the clustering process forms the next class by finding a maximal complete sub-graph of the complete weighted graph containing vertices $D_{r+1}, D_{r+2}, \ldots, D_n$. A maximal complete sub-graph is defined as follows:

Definition 6.6 A weighted complete sub-graph g of a complete weighted graph G is maximal at the similarity level δ if the following conditions are satisfied: (i) The weight of every edge of g is greater than or equal to δ. (ii) The addition of one more vertex (i.e., a database) to g leads to the addition of at least one edge to g having weight less than δ.

We need to find out a maximal weighted complete sub-graph of the complete weighted graph of the remaining vertices to form the next class. This process continues till all the vertices have been clustered. A clustering of databases is defined as follows:

Definition 6.7 Let D be a set of databases. Let $\pi_k^\delta\ (D, \alpha)$ be a clustering of databases in D at the similarity level δ under the similarity measure $simi_k$. Then, $\pi_k^\delta\ (D, \alpha) = \{X : X \in \rho(D)$, and X is a $class_\kappa^\delta\ (D, \alpha)\}$, where $\rho(D)$ is the power set of D, $(k = 1, 2)$.

During the clustering process we may like to impose a restriction that each database belongs to at least one class. This restriction makes a clustering complete. We define a complete clustering as follows:

Definition 6.8 Let D be a set of databases. Let $\pi_k^\delta(D, \alpha) = \{C_k^\delta(D, \alpha), C_{k,2}^\delta(D, \alpha), \ldots, C_{k,m}^\delta(D, \alpha)\}$, where $C_{k,i}^\delta(D, \alpha)$ is the i-th class of π_k^δ, $i = 1, 2, \ldots, m$. π_k^δ is complete, if $\cup_{i=1}^m \mathrm{C}_{\mathrm{k,i}}^\delta\ D$, $(k = 1, 2)$.

In complete clustering, two classes may have a common database. We may be interested in forming clustering of mutually exclusive classes. We define mutually exclusive clustering as follows:

Definition 6.9 Let D be a set of databases. Let $\pi_k^\delta\ (D, \alpha) = \{C_{k,1}^\delta(D, \alpha), C_{k,2}^\delta(D, \alpha), \ldots, C_{k,m}^\delta(D, \alpha)\}$, where $C_{k,i}^\delta(D, \alpha)$ is the i-th class of π_k^δ, $i = 1, 2, \ldots, m$. π_k^δ is mutually exclusive if $C_{k,i}^\delta(D, \alpha) \cap C_{k,j}^\delta\ (D, \alpha) = \phi$, for $i \neq j$, $1 \leq i, j \leq m$, $(k = 1, 2)$.

We may be interested in realizing such a mutually exclusive and complete clustering. Here we have

Definition 6.10 Let D be a set of databases. Also, let $\pi_k^\delta(D, \alpha)$ be a clustering of databases in D at the similarity level δ under the similarity measure $simi_k$. If $\pi_k^\delta(D, \alpha)$ is a mutually exclusive and complete clustering then it is called a partition ($k = 1, 2$).

Definition 6.11 Let D be a set of databases. Also let $\pi_k^\delta(D, \alpha)$ be a partition of D at the similarity level δ under the similarity measure $simi_k$. $\pi_k^\delta(D, \alpha)$ is called a non-trivial partition if $1 < |\pi_k^\delta| < n$ ($k = 1, 2$).

A clustering does not necessarily lead to a partition. In the following example, we wish to find partitions (if they exist) of a set of databases.

Example 6.4 With reference to Example 6.2, consider the set of databases $D = \{D_1, D_2, \ldots, D_7\}$. The corresponding DSM_2 is given as follows.

$$DSM_2(D, 0.35) = \begin{bmatrix} 1.0 & 0.780 & 0.539 & 0.0 & 0.0 & 0.0 & 0.0 \\ 0.780 & 1.00 & 0.636 & 0.0 & 0.0 & 0.0 & 0.0 \\ 0.539 & 0.636 & 1.0 & 0.061 & 0.0 & 0.0 & 0.0 \\ 0.0 & 0.0 & 0.061 & 1.0 & 0.063 & 0.065 & 0.087 \\ 0.0 & 0.0 & 0.0 & 0.063 & 1.0 & 0.641 & 0.353 \\ 0.0 & 0.0 & 0.0 & 0.065 & 0.641 & 1.0 & 0.444 \\ 0.0 & 0.0 & 0.0 & 0.087 & 0.353 & 0.444 & 1.0 \end{bmatrix},$$

We arrange all non-zero and distinct $DSM_2^{i,j}(D, 0.35)$ values in non-increasing order, for $1 \le i < j \le 7$. The arranged similarity values are given as follows: 0.780, 0.641, 0.636, 0.539, 0.444, 0.353, 0.087, 0.065, 0.063, 0.061. We obtain many non-trivial partitions formed at different similarity levels. At the similarity levels equal to 0.780, 0.641, 0.539, and 0.353, we get non-trivial partitions as $\pi_2^{0.780} = \{ \{D_1, D_2\}, \{D_3\}, \{D_4\}, \{D_5\}, \{D_6\}, \{D_7\} \}$, $\pi_2^{0.641} = \{ \{D_1, D_2\}, \{D_3\}, \{D_4\}, \{D_5, D_6\}, \{D_7\} \}$, $\pi_2^{0.539} = \{ \{D_1, D_2, D_3\}, \{D_4\}, \{D_5, D_6\}, \{D_7\} \}$, and $\pi_2^{0.353} = \{ \{D_1, D_2, D_3\}, \{D_4\}, \{D_5, D_6, D_7\} \}$, respectively.

Our *BestDatabasePartition* algorithm (as presented in Section 6.4.1) is based on binary similarity matrix (*BSM*). We derive binary similarity matrix BSM_k from the corresponding DSM_k ($k = 1, 2$). BSM_k is defined as follows:

Definition 6.12 The (i, j)-th element of the binary similarity matrix BSM_k at the similarity level δ using the similarity measure $simi_k$ is defined as follows.

$$BSM_k^{i,j}(D, \alpha, \delta) = \begin{cases} 1, \text{if } simi_k\,(D_i\,, D_j, \alpha) \ge \delta \\ 0, \text{ otherwise} \end{cases}, \quad \text{for} \quad i,j = 1, 2, \ldots, n(\,k = 1, 2).$$

We take an example of BSM_2 and observe the distribution of 0s and 1s.

Example 6.5 With reference to Example 6.4, the *BSM*$_2$ at the similarity level 0.353 is given below.

$$BSM2(D, 0.35, 0.353) = \begin{bmatrix} 1\,1\,1\,0\,0\,0\,0 \\ 1\,1\,1\,0\,0\,0\,0 \\ 1\,1\,1\,0\,0\,0\,0 \\ 0\,0\,0\,1\,0\,0\,0 \\ 0\,0\,0\,0\,1\,1\,1 \\ 0\,0\,0\,0\,1\,1\,1 \\ 0\,0\,0\,0\,1\,1\,1 \end{bmatrix}, \text{where } D = \{D_1, D_2, \ldots, D_7\}.$$

There may exist two the same partitions at two distinct similarity levels. Two partitions are distinct if they are not the same. In the following, we define two same partitions at two distinct similarity levels.

Definition 6.13 Let D be a set of databases. Let $C \subseteq D$, and $C \neq \phi$. Two partitions $\pi_k^{\delta_1}(D, \alpha)$ and $\pi_k^{\delta_2}(D, \alpha)$ are the same, if the following statement is true: $C \in \pi_k^{\delta_1}$ if and only if $C \in \pi_k^{\delta_2}$, for $\delta_1 \neq \delta_2$.

We would like to enumerate the maximum number of possible distinct partitions. In Theorem 6.5, we find the maximum number of possible distinct partitions of a set of databases (Adhikari and Rao 2008).

Theorem 6.5 *Let D be a set of databases. Let m be the number of distinct non-zero similarity values in the upper triangle of DSM*$_2$*. Then the number of distinct partitions is less than or equal to m.*

Proof We arrange the non-zero similarity values of the upper triangle of *DSM*$_2$ in non-increasing order. Let $\delta_1, \delta_2, \ldots, \delta_m$ be m non-zero ordered similarity values. Let δ_i , δ_{i+1} be two consecutive similarity values in the sequence of non-increasing similarity values. Let $x, y \in [\delta_i , \delta_{i+1})$, for some $i = 1, 2, \ldots, m$, where $\delta_{m+1} = 0$. Then $BSM_2(D, \alpha, x) = BSM_2(D, \alpha, y)$. Thus, there exists at the most one distinct non-trivial partition in the interval $[\delta_i , \delta_{i+1})$, for $i = 1, 2, \ldots, m$. We have m such semi-closed intervals $[\delta_i , \delta_{i+1})$, $i = 1, 2, \ldots, m$. The theorem follows.

For the purpose of finding partitions of the input databases, we first design a simple algorithm that uses the apriori property (Agrawal et al. 1993). The similarity values considered here are based on the similarity measure *simi*$_2$. Initially, we have n database classes, where n is the number of databases. At this time, each class contains a single database object. These classes are assumed at level 1. Based on the classes at level 1, we construct database classes at level 2. At level 1, we assume that the i-th class contains database D_i, $i = 1, 2, \ldots, n$. i-th class and j-th class of level 1 could be merged if $simi_2(D_i, D_j) \geq \delta$, where δ is the user-defined level of similarity. We proceed further until no more classes could be generated and no more levels could be generated. The algorithm (Adhikari and Rao 2008) is presented below.

Algorithm 6.1 Find partitions (if they exist) of a set of databases using apriori property.

```
procedure AprioriDatabaseClustering (n, DSM2)
Input: n, DSM2
n: number of databases
DSM2: database similarity matrix
Output: Partitions (if they exist) of input databases
01:  sort all the non-zero values that exist in the upper triangle of DSM2 in non-increasing
02:  order into an array called simValues; let the number of non-zero values be m;
03:  let k = 1; let simValues(m +1) = 0; let delta = simValues(k);
04:  while (delta > 0) do
05:    construct n classes, where each class contains a single database; // level: 1
06:    repeat line 7 until no more level could be generated;
07:      construct all possible classes at level (i +1) using lines 8-10; // level: i +1
08:        let A and B be two classes at the i-th level such that |A ∩ B| = i -1;
09:        let a∈ (A-B), and b∈ (B-A);
10:        if ≥ δ then construct a new class A ∪ B; end if
11:    repeat line 12 from top level to level 1;
12:      for each class at the current level do
13:      if all databases of the current class are not included a class generated earlier then
14:        generate the current class;
15:      end if
16:      end for
17:      if the current clustering is a partition then store it; end if
18:      increase k by 1; let delta = simValues(k);
19:  end while
20:  display all the partitions;
end procedure
```

Lines 1–2 take $O(m \times \log(m))$ time to sort m data. While-loop at line 4 executes m times. Line 5 takes $O(n)$ time. Initially (at line 5), n classes are constructed. At the first iteration of line 6, the maximum number of classes generated is $^{n}C_2$. At the second iteration, the maximum number of classes generated is $^{n}C_3$. Lastly, at the (n–1)-th iteration, the maximum number of classes generated is $^{n}C_n$. Thus, the maximum number of possible classes is $O\left(\sum_{i=1}^{n} {}^{n}C_i\right)$, i.e., $O(2^n)$. Let p be the average size of a class. Line 8 takes $O(p)$ time. Also, line 11 takes $O(2^n)$ time, since the maximum number of possible classes is $O(2^n)$. Thus, the time complexity of lines 4–19 is $O(m \times p \times 2^n)$. The line 20 takes time $O(m \times n)$, since the maximum number of partitions is m. Thus, the time complexity of the procedure *AprioriDatabaseClustering* is *maximum* $\{O(m \times \log(m), O(m \times p \times 2^n), O(m \times n)\}$, i.e., $O(m \times p \times 2^n)$, since $p \times 2^n > 2^n > n^2 > m > \log_2(m)$, for $p > 1$ and $n > 4$. The *AprioriDatabaseClustering* algorithm generates all possible classes level-wise. It is a simple but not an efficient clustering technique, since the time-complexity of the algorithm is an exponential function of n.

In Theorems 6.6–6.9, we discuss some properties of BSM_2.

Theorem 6.6 *Let* $D = \{D_1, D_2, \ldots, D_n\}$. *Let* $\pi_2^{\delta}(D, \alpha)$ *be a clustering of databases in* D *at the similarity level* δ.π_2^{δ} *is a partition if and only if the corresponding* BSM_2

gets transformed into the following form by inter-changing jointly a row and the corresponding column with another row and the corresponding column.

$$\begin{bmatrix} U_1 & 0 & \ldots & 0 \\ 0 & U_2 & \ldots & 0 \\ \ldots & \ldots & \ldots & \ldots \\ 0 & 0 & \ldots & U_m \end{bmatrix},\ U_i \text{ is a matrix of size } n_i \times n_i, \text{ containing all elements as } 1,$$

where $\sum_{i=1}^{m} n_i = n$, $|\pi_2^\delta| = m$.

Proof Let { $D_1^i, D_2^i, \ldots, D_{n_i}^i$ } be the i-th database class of the partition at the similarity level δ, $i = 1, 2, \ldots, m$. The row referring to D_j^i of BSM_2 corresponds to a unique combination of 0s and 1s, $j = 1, 2, \ldots, n_i$. Similarly, the column corresponding to D_j^i of BSM_2 results as a unique combination of 0s and 1s, $j = 1, 2, \ldots, n_i$. All such n_i rows and columns may not be initially consecutive, $i = 1, 2, \ldots, m$. We keep these n_i rows and columns consecutive, $i = 1, 2, \ldots, m$. Initially, we keep n_1 rows and the corresponding columns of the first database class to be consecutive. Then, we keep n_2 rows and the corresponding columns of the second database class consecutively and so on. In general, to fix the matrix U_i at the proper position, we interchange jointly $\left(\sum_1^{i-1} n_j + k\right)$-th row and $\left(\sum_1^{i-1} n_j + k\right)$-th column with D_k^i-th row and D_k^i-th column of BSM_2, $1 \le k \le n_i$, $i = 1, 2, \ldots, m$.

Referring to BSM_2 in Example 6.5, we apply Theorem 6.6, and conclude that a partition exists at similarity level of 0.353.

Theorem 6.7 *Let D be a set of databases. Let $\pi_2^\delta(D, \alpha)$ be a clustering of databases in D at the similarity level δ. Let { $D_1^i, D_2^i, \ldots, D_{n_i}^i$ } be the i-th database class of $\pi_2^\delta(D, \alpha)$. Then D_k^i-th row (or, D_k^i-th column) of BSM_2 contains n_i 1s, $k = 1, 2, \ldots, n_i$, $i = 1, 2, \ldots, |\pi_2^\delta|$.*

Proof If possible, let D_k^i-th row or, D_k^i-th column has (n_i +1) 1s. Then D_k^i-th database would belong to two database classes. It contradicts the mutual exclusiveness of classes of a partition. If possible, let D_k^i-th row or, D_k^i-th column contains (n_i-1) 1s. It contradicts the fact that $BSM_2^{D_j^i, D_k^i} = 1$, $j = 1, 2, \ldots, n_i$ and $j \neq k$.

Theorem 6.8 *Let D be a set of databases. Let $\pi_2^\delta(D, \alpha)$ be a clustering of databases in D at the similarity level δ. Then, the rank of the corresponding BSM_2 is $|\pi_2^\delta|$.*

Proof Let { $D_1^i, D_2^i, \ldots, D_{n_i}^i$ } be the i-th database class of π_2^δ. Then,

$$BSM_2^{D_j^i, D_k^i}(\mathrm{D}, \alpha) = \begin{cases} 1, & \text{for } D_j^i, D_k^i \in \{D_1^i, D_2^i, \ldots, D_{n_i}^i\} \\ 0, & \text{for } D_j^i \in \{D_1^i, D_2^i, \ldots, D_{n_i}^i\} \text{ and } D_k^i \notin \{D_1^i, D_2^i, \ldots, D_{n_i}^i\} \\ 0, & \text{for } D_j^i \notin \{D_1^i, D_2^i, \ldots, D_{n_i}^i\} \text{ and } D_k^i \in \{D_1^i, D_2^i, \ldots, D_{n_i}^i\} \end{cases}$$

The row corresponding to D_j^i of BSM_2 corresponds to a unique combination of 0s and 1s, for $j = 1, 2, \ldots, n_i$. So, all the rows of BSM_2 are divided into $|\pi_2^\delta|$ groups such that all the rows in a group correspond to a unique combination of 0s and 1s. Thus, BSM_2 has $|\pi_2^\delta|$ independent rows.

Theorem 6.9 *Let $D = \{D_1, D_2, \ldots, D_n\}$. At a given value of the triplet (D, α, δ), there exists at the most one partition of D.*

Proof At a given value of the pair (D, α), the element $DSM_2^{i,j}$ is unique, $i, j = 1, 2, \ldots, n$. Thus at a given value of the tuple (D, α, δ) the element $BSM_2^{i,j}$ is unique, for $i, j = 1, 2, \ldots, n$. There exists a partition if the BSM_2 gets transformed into a specific form (as outlined in Theorem 6.6), by jointly interchanging a row and the corresponding column with another row and the corresponding column. Hence, the theorem follows.

6.3.1 Finding the Best Non-trivial Partition

Now we get back to Example 6.4. We observed that at different similarity levels there may exist different partitions. We have observed the existence of four non-trivial partitions. We would like to find the best partition among these partitions. The best partition is based on the principle of maximizing the intra-class similarity and maximizing the inter-class distance. The intra-class similarity and inter-class distance are defined as follows:

Definition 6.14 The intra-class similarity *intra-sim* of a partition π at the similarity level δ using the similarity measure $simi_2$ is defined as follows:

$$intra\text{-}sim(\pi_2^\delta) = \sum_{C \in \pi_2^\delta} \sum_{D_i, D_j \in C; i<j} simi_2(D_i, D_j, \alpha).$$

Definition 6.15 The inter-class distance *inter-dist* of a partition π at the similarity level δ using the similarity measure $simi_2$ is defined as follows:

$$inter\text{-}dist(\pi_2^\delta) = \sum_{C_p, C_q \in \pi_2^\delta; p<q} \sum_{D_i \in C_p; D_j \in C_q; i<j} dist_2(D_i, D_j, \alpha).$$

The best partition among a set of partitions is selected on the basis of goodness value of a partition. The goodness measure itself, *goodness,* of a partition is defined as follows:

Definition 6.16 The goodness of a partition π at similarity level δ using the similarity measure $simi_2$ is expressed as follows:

$$goodness(\pi_2^\delta) = intra\text{-}sim(\pi_2^\delta) + inter\text{-}dist(\pi_2^\delta) - |\pi_2^\delta|,$$

where $|\pi_2^\delta|$ is the number classes in π.

Note that we have subtracted the term $|\pi_2^\delta|$ from the sum of intra-class similarity and inter-class distance to remove the bias of goodness value of a partition. The higher the value of *goodness*, the better is the corresponding partition. Now, we partition the set of databases D using the proposed goodness measure.

Example 6.6 Continuing Example 6.4, we calculate the goodness value of each of the non-trivial partitions using $simi_2$ as follows:

$$
\begin{aligned}
\textit{intra-sim}(\pi_2^{0.353}) &= 3.185,\ \textit{inter-dist}(\pi_2^{0.353}) \\
&= 15.276,\ |\pi_2^{0.353}| = 3, \text{and}\ \textit{goodness}(\pi_2^{0.353}) = 15.461. \\
\textit{intra-sim}(\pi_2^{0.539}) &= 2.596,\ \textit{inter-dist}(\pi_2^{0.539}) \\
&= 16.666,\ |\pi_2^{0.539}| = 4, \text{and}\ \textit{goodness}(\pi_2^{0.539}) = 15.262. \\
\textit{intra-sim}(\pi_2^{0.641}) &= 1.421,\ \textit{inter-dist}(\pi_2^{0.641}) \\
&= 17.491,\ |\pi_2^{0.641}| = 5, \text{and}\ \textit{goodness}(\pi_2^{0.641}) = 13.912. \\
\textit{intra-sim}(\pi_2^{0.780}) &= 0.780,\ \textit{inter-dist}(\pi_2^{0.780}) \\
&= 17.118,\ |\pi_2^{0.780}| = 6, \text{and}\ \textit{goodness}(\pi_2^{0.780}) = 11.898.
\end{aligned}
$$

The goodness value corresponding to the partition $\pi_2^{0.353}$ attains the maximal value. The partition $\pi_2^{0.353} = \{\ \{D_1, D_2, D_3\}, \{D_4\}, \{D_5, D_6, D_7\}\ \}$ is the best among all the non-trivial partitions. Let us look back into the databases presented in Example 6.2. We find that the partition $\pi_2^{0.353}$ matches the best the ground reality among the partitions reported.

We present an algorithm (Adhikari and Rao 2008) for finding the best non-trivial partition of a set of databases.

Algorithm 6.2 Find the best non-trivial partition (if it exists) of a set of databases.
procedure *BestDatabasePartition* (n, DSM_2)

Input: n, DSM_2
n: number of databases
DSM_2: database similarity matrix

Output: The best partition (if it exists) of input databases
sort all the non-zero values that exist in the upper triangle of DSM_2 in non-increasing order into an array called *simValues*; **let** the number of non-zero values be m;
let $k = 1$; **let** $simValues(m + 1) = 0$; **let** $delta = simValues(k)$;
while ($delta > 0$) **do**
  for $i = 1$ to n **do** $class(i) = 0$; **end for**
  construct the BSM_2 at current level of the similarity *delta*;
  let $currentClass = 1$; **let** $currentRow = 1$; **let** $class(1) = currentClass$;

```
08:      for col = (currentRow + 1) to n do
09:        if (BSM2^{currentRow, col} = 1) then
10:          if (class(col) = 0) then class(col) = currentClass;
11:          else if (class(col) ≠ currentClass) then go to line 24; end if
12:          end if
13:          end if
14:      end for
15:      let i = 1; let class(n +1) = 0;
16:      while (class(i) ≠ 0) do increase i by 1; end while
17:      if (i = n +1) then
18:        store the content of array class and current similarity level delta;
19:      else
20:        increase currentRow by 1;
21:        if (class(currentRow) = 0) then increase currentClass by 1; end if
22:        go to line 8;
23:      end if
24:      increase k by 1; let delta = simValues(k);
25:    end while
26:    for each non-trivial partition do
27:      calculate the goodness value of the current partition;
28:    end for
29:    return the partition whose goodness value is the maximum;
end procedure
```

Initially, we have sorted all non-zero values in the upper triangle of DSM_2 in non-increasing order. The algorithm checks the existence of a partition starting with the maximum of all the similarity values. At line 5, we initialize the class label of each database to 0. The algorithm starts forming a class with D_1 (the first database) as the variable *currentRow* is initialized with 1. Also, class label starts with 1 as the variable *currentClass* is initialized with 1. Lines 8–14 are used to check the similarity of $D_{\text{currentRow}}$ with other databases. If the condition at line 9 is true then databases $D_{\text{currentRow}}$ and D_{col} are similar. At line 10, D_{col} is put in the *currentClass* if it is still unlabelled. If D_{col} is already labeled with a class label not equal to current class label then D_{col} get another label. Thus, partition does not exist at the current similarity level.

Operations in Line 1 take $O(m \times \log(m)$ time. Line 3 repeats m times. Line 6 constructs BSM_2 in $O(n^2)$ time as the order of BSM_2 is $n \times n$. Each of lines 5 and 16 takes $O(n)$ time. For-loop positioned at line 8, repeats maximum n times. Line 18 takes $O(n)$ time, since the time required to store a partition is $O(n)$. Thus, the time complexity of lines 4–25 is $O(m \times n^2)$. Therefore, the time complexity of the procedure *best-database-partition* is *maximum* { $O(m \times \log(m)$, $O(m \times n^2)$ }, i.e., $O(m \times n^2)$, since $n^2 > m > \log_2(m)$.

The drawback of *BestClassification* (Wu et al. 2005b) algorithm is that the step value for assigning the next similarity level has to be user-defined. Thus, the method might fail to find the exact similarity level at which a partition exists.

BestDatabasePartition algorithm reports the exact similarity level at which a partition exists. Also, the algorithm works faster, since it is required to check for the existence of partitions only at m similarity levels. Li et al. (2009) have recently proposed *BestCompleteClass* algorithm for partitioning a set of databases. But, the *BestCompleteClass* algorithm has followed the strategies that we have already reported in the *BestDatabasePartition* algorithm. In Theorem 6.10, we prove the correctness of the proposed algorithm.

Theorem 6.10 *Algorithm BestDatabasePartition works correctly.*
Proof Let $D = \{D_1, D_2, \ldots, D_n\}$. Let there are m distinct non-zero similarity values in the upper triangle of DSM_2. Using Theorem 6.5, one could conclude that the maximum number of partitions of D is m at a given value of pair (D, α). While-loop at line 4 checks for the existence of partitions at m similarity levels. At each similarity level, we get a new BSM_2. The existence of a partition is determined from the BSM_2. We have an array *class* that stores the class label given to each database under the current level of similarity. In a partition, each database has a unique class label. The existence of a partition is checked based on the principle that every database receives a unique class label. As soon as we find that a labeled database receives another class label, we conclude that a partition does not exist at the current level of similarity *delta* (line 11). Initially, we put the class label 0 to all databases using line 5. Then, we start from the row 1 of BSM_2 that corresponds to database D_1. Thus, D_1 is kept in the first database class. If there is a 1 in the j-th column of BSM_2, then we put class label of D_j as 1 using line 10. We find a database D_i that has not been clustered yet using lines 15–16. Then, we start at row i of BSM_2. If there is a 1 in the j-th column of row i, then we put database D_j in the current class. Thus, the algorithm *BestDatabasePartition* works correctly. •

6.3.2 Efficiency of Clustering Technique

The proposed clustering algorithm is based on the similarity measure $simi_2$. The same similarity measure $simi_2$ is based on the supports of the frequent itemsets in databases. If we vary the value of α then the number of frequent itemsets in a database changes. The accuracy of similarity between two databases increases as the number of frequent itemsets increases. Therefore, a clustering process would be more accurate for lower values of α. The frequent itemsets participate in the clustering process is limited by main memory. If we can store more frequent itemsets in main memory then $simi_2$ could determine similarity between two databases more accurately. Thus, the clustering process would be more accurate. This limitation begs for a space efficient representation of the frequent itemsets in main memory. For this purpose, we propose a coding that efficiently represent frequent itemsets. The coding allows more frequent itemsets to participate in determining the similarity between two databases.

6.3.2.1 Space Efficient Representation of Frequent Itemsets in Different Databases

In this technique, we represent each frequent itemset using a bit vector. Each frequent itemset has three components: database identification, frequent itemset, and support. Let the number of databases be n. There exists an integer p such that $2^{p-1} < n \leq 2^p$. Then p bits are enough to represent a database. Let k be the number of digits after the decimal point to represent support. Support value 1.0 could be represented as 0.99999, for $k = 5$. If we represent the support s as an integer d containing of k digits then $s = d \times 10^{-\mathrm{k}}$. The number digits required to represent a decimal number could be obtained by Theorem 5.3. The proposed coding is described with the help of Example 6.7.

Example 6.7 We refer again to Example 6.2. The frequent itemsets sorted in non-increasing order with regard to the number of extractions are given as follows:

(h, 4), (a, 3), (ac, 3), (c, 3), (hi,3), (i, 3), (j, 2), (e, 2), (ij, 2), (ae, 1), (d, 1), (df, 1), (ef, 1), (f, 1), (fh, 1), (g, 1), (gi, 1).

(X, μ) denotes itemset X having number of extractions equal to μ. We code the frequent itemsets of the above table from left to right. The frequent itemsets are coded using a technique similar to Huffman coding (Huffman 1952). We attach code 0 to itemset h, 1 to itemset a, 00 to itemset ac, 01 to itemset c, etc. Itemset h gets a code of minimal length, since it has been extracted maximum number of times. We call this coding as itemset (IS) coding. It is a lossless coding (Sayood 2000). IS coding and Huffman coding are not the same, in the sense that an IS code may be a prefix of another IS code. Coded itemsets are given as follows:

(h, 0), (a, 1), (ac, 00), (c, 01), (hi, 10), (i, 11), (j, 000), (e, 001), (ij, 010), (ae, 011), (d, 100), (df, 101), (ef, 110), (f, 111), (fh, 0000), (g, 0010), (gi, 0011). Here (X, ν) denotes itemset X having IS code ν.

6.3.2.2 Efficiency of IS Coding

Using the above representation of the frequent itemsets, one could store more frequent itemsets in the main memory during the clustering process. This enhances the efficiency of the clustering process.

Definition 6.17 Let there are n databases $D_1, D_2, \ldots, D_n$. Let $S^T\left(\cup_{i=1}^{n} FIS\,(D_i)\right)$ be the amount of storage space (in bits) required to represent $\cup_{i=1}^{n} FIS\,(D_i)$ by a technique T. Let $S_{min}\left(\cup_{i=1}^{n} FIS\,(D_i)\right)$ be the minimum amount of storage space (in bits) required to represent $\cup_{i=1}^{n} FIS\,(D_i)$. Let τ, κ, and λ denote a clustering algorithm, similarity measure, and computing resource under consideration, respectively. Let Γ be the set of all frequent itemset representation techniques. We define efficiency of a frequent itemset representation technique T at a given value of triplet (τ, κ, λ) as follows:

$$\varepsilon(\mathrm{T}|\tau, \kappa\ , \lambda) = S_{min}\left(\cup_{i=1}^{n} FIS\,(D_i)\right)/S^T\left(\cup_{i=1}^{n} FIS\,(D_i)\right), \text{ for T} \in \Gamma.$$

One could store an itemset conveniently using the following components: database identification, items in the itemset, and support. Database identification, an item and a support could be stored as a short integer, an integer and a real type data, respectively. A typical compiler represents a short integer, an integer and a real number using 2, 4 and 8 bytes, respectively. Thus, a frequent itemset of size 2 could consume $(2 + 2 \times 4 + 8) \times 8$ bits, i.e. 144 bits. An itemset representation may have an overhead of indexing frequent itemsets. Let *OI*(*T*) be the overhead of indexing coded frequent itemsets using technique *T*.

Theorem 6.11 *IS coding stores a set of frequent itemsets using minimum storage space, if OI(IS coding)* $\le$ *OI(T), for* $T \in \Gamma$.

Proof A frequent itemset has three components, viz., database identification, itemset, and support. Let the number of databases be n. Then $2^{p-1} < n \le 2^p$, for an integer p. We need minimum p bits to represent a database identifier. The representation of database identification is independent of the corresponding frequent itemsets. If we keep k digits to store a support then $\lceil k \times \log_2 10 \rceil$ binary digits are needed to represent a support (as mentioned in Theorem 5.3). Thus, the representation of support becomes independent of the other components of the frequent itemset. Also, the sum of all IS codes is the minimum because of the way they are constructed. Thus, the space used by the IS coding for representing a set of frequent itemsets attains the minimum.

Thus, the efficiency of a frequent itemset representation technique T could be expressed as follows:

$$\varepsilon(\mathrm{T}|\tau,\kappa,\lambda) = S^{\text{IS coding}}\left(\cup_{i=1}^{n} FIS\,(D_i)\right)/S^{T}\left(\cup_{i=1}^{n} FIS\,(D_i)\right), \quad \text{provided } OI(\text{IS coding}) \le OI(T), \text{ for } T \in \Gamma. \tag{6.15}$$

If the condition in (6.15) is satisfied, then the IS coding performs better than any other techniques. If the condition in (6.15) is not satisfied, then the IS coding performs better than any other techniques in almost all cases. The following corollary is derived from Theorem 6.11.

Corollary 6.1 *Efficiency of IS coding attains maximum, if OI(IS coding)* $\le$ *OI(T), for* $T \in \Gamma$.

Proof $\varepsilon(\text{IS coding} \mid \tau, \kappa, \lambda) = 1.0$.

The IS coding maintains an index table to decode/search a frequent itemset. In the following example, we compute the amount of space required to represent the frequent itemsets using an ordinary method and the IS coding.

Example 6.8 With reference to Example 6.7, there are 33 frequent itemsets in different databases. Among them, there are 20 itemsets of size 1 and 13 itemsets of size 2. An ordinary method could use $(112 \times 20 + 144 \times 13) = 4{,}112$ bits. The amount of space required to represent frequent itemsets in seven databases using

IS coding is equal to $P + Q$ bits, where P is the amount of space required to store frequent itemsets, and Q is the amount of space required to maintain the index table. Since there are seven databases, we need 3 bits to identify a database. The amount of memory required to represent the database identification for 33 frequent itemsets is equal to 33×3 bits $= 99$ bits. Suppose we keep 5 digits after the decimal point for a support. Thus, $\lceil 5 \times \log_2(10) \rceil$ bits, i.e., 17 bits are required to represent a support. The amount of memory required to represent the supports of 33 frequent itemsets is equal to 33×17 bits $= 561$ bits. Let the number of items be 10,000. Therefore, 14 bits are required to identify an item. The amount of storage space would require for itemsets h and ac are 14 and 28 bits respectively. To represent 33 frequent itemsets, we need $(20 \times 14 + 13 \times 28)$ bits $= 644$ bits. Thus, $P = (99 + 561 + 644)$ bits $= 1{,}304$ bits. There are 17 frequent itemsets in the index table. Using IS coding, 17 frequent itemsets consume 46 bits. To represent 17 frequent itemsets, we need $14 \times 9 + 28 \times 8$ bits $= 350$ bits. Thus, $Q = 350 + 46$ bits $= 396$ bits. The total amount of memory space required (including the overhead of indexing) to represent frequent itemsets in 7 databases using IS coding is equal to $P + Q$ bits, i.e., 1,700 bits. The amount of space saving in compared to an ordinary method is equal to 2,412 bits, i.e., 58.66% approximately. A technique without optimization (TWO) may not maintain index table separately. In this case, OI(TWO) = 0. In spite of that, IS coding performs better than a TWO in most of the cases.

Finally, we claim that our clustering technique is more accurate. There are two reasons for this claim: (i) We propose more appropriate measures of similarity than the existing ones. We have observed that the similarity between two databases based on items might not be appropriate. The proposed measures are based on the similarity between transactions of two databases. As a consequence the similarity between two databases is estimated more accurately. (ii) Also, the proposed IS coding enables us to mine local databases further at a lower level of α to accommodate more frequent itemsets in main memory. As a result, more frequent itemsets could participate in the clustering process.

6.4 Experiments

We have carried out a number of experiments to study the effectiveness of our approach. We present experimental results using two synthetic databases, and one real database. The synthetic databases *T10I4D100K* (Frequent itemset mining dataset repository 2004) and *T40I10D100K* (Frequent itemset mining dataset repository 2004) have been generated using synthetic database generator from IBM Almaden Quest research group. The real database *BMS-Web-Wiew-1* could be found at the KDD CUP 2000 repository (KDD CUP 2000). Let *NT*, *ALT*, *AFI*, and *NI* denote the number of transactions, the average length of a transaction, the average frequency of an item, and the number of items in the database (*DB*), respectively.

Each of the above databases is divided into 10 databases for the purpose of carrying out experiments. The databases obtained from *T10I4D100K*, and *T40I10D100K* are named T_{1j}, and T_{4j}, respectively, $j = 0, 1, \ldots, 9$. The databases obtained from *BMS-Web-Wiew-1* are named B_{1j}, $j = 0, 1, \ldots, 9$. The databases T_{ij} and B_{1j} are called input databases, for $i = 1, 4$, and $j = 0, 1, \ldots, 9$. Some characteristics of these input databases are presented in the Table 6.1.

At a given value of α, there may exist many partitions. Partitions of the set of input databases are presented in Table 6.2. If we vary the value of α, the set of frequent itemsets in a database varies. Apparently, the similarity between a pair of databases changes over the change of α.

At a lower value of α, more frequent itemsets are reported from a database and hence the database is represented more correctly by its frequent itemsets. We obtain a more accurate value of similarity between a pair of databases. Thus, the partition generated at a smaller value of α would be more correct. In Tables 6.3 and 6.4, we have presented best partitions of a set of databases obtained for different values of α. So, the best partition of a set of databases may change over the change of α.

Thus, a partition may not remain the same over the change of α. But, we have observed a general tendency that the databases show more similarity over larger values of α. As the value of α becomes smaller, more frequent itemsets are reported from a database, and databases become more dissimilar.

In Fig. 6.3, we have shown how the execution time of an experiment increases as the number databases increases. The execution time increases faster as we increase input databases from database T_1. The reason is that the size of each local database obtained from T_1 is larger than that of T_4 and B_1.

The number of frequent itemsets decreases as the value of α increases. Thus, the execution time of an experiment decreases as α increases. We observe this phenomenon in Figs. 6.4 and 6.5.

Table 6.1 Input database characteristics

DB	*NT*	*ALT*	*AFI*	*NI*	*DB*	*NT*	*ALT*	*AFI*	*NI*
T_{10}	10,000	11.06	127.66	866	T_{15}	10,000	11.14	128.63	866
T_{11}	10,000	11.13	128.41	867	T_{16}	10,000	11.11	128.56	864
T_{12}	10,000	11.07	127.65	867	T_{17}	10,000	11.10	128.45	864
T_{13}	10,000	11.12	128.44	866	T_{18}	10,000	11.08	128.56	862
T_{14}	10,000	11.14	128.75	865	T_{19}	10,000	11.08	128.11	865
T_{40}	10,000	40.57	431.57	940	T_{45}	10,000	40.51	430.46	941
T_{41}	10,000	40.58	432.19	939	T_{46}	10,000	40.74	433.44	940
T_{42}	10,000	40.63	431.79	941	T_{47}	10,000	40.62	431.71	941
T_{43}	10,000	40.63	431.74	941	T_{48}	10,000	40.53	431.15	940
T_{44}	10,000	40.66	432.56	940	T_{49}	10,000	40.58	432.16	939
B_{10}	14,000	2.00	14.94	1,874	B_{15}	14,000	2.00	280.00	100
B_{11}	14,000	2.00	280.00	100	B_{16}	14,000	2.00	280.00	100
B_{12}	14,000	2.00	280.00	100	B_{17}	14,000	2.00	280.00	100
B_{13}	14,000	2.00	280.00	100	B_{18}	14,000	2.00	280.00	100
B_{14}	14,000	2.00	280.00	100	B_{19}	23,639	2.00	472.78	100

Table 6.2 Partitions of the input databases for a given value of α

Databases	α	Non-trivial distinct partition (π)	δ	*Goodness* (π)
$\{T_{10}, \ldots, T_{19}\}$	0.03	$\{\{T_{10}\},\{T_{11}\},\{T_{12}\},\{T_{13}\},\{T_{14},T_{18}\},\{T_{15}\},\{T_{16}\},\{T_{17}\},\{T_{19}\}\}$	0.881	0.01
$\{T_{40}, \ldots, T_{49}\}$	0.1	$\{\{T_{40}\},\{T_{41}, T_{45}\},\{T_{42}\},\{T_{43}\},\{T_{44}\},\{T_{46}\},\{T_{47}\},\{T_{48}\},\{T_{49}\}\}$	0.950	−3.98
		$\{\{T_{40}\},\{T_{41},T_{45}\},\{T_{42}\},\{T_{43}\},\{T_{44}\},\{T_{46}\},\{T_{47}\},\{T_{48},T_{49}\}\}$	0.943	11.72
		$\{\{T_{40}\},\{T_{41}, T_{43}, T_{45}\},\{T_{42}\},\{T_{44}\},\{T_{46}\},\{T_{47}\},\{T_{48}, T_{49}\}\}$	0.942	24.21
$\{B_{10}, \ldots, B_{19}\}$	0.009	$\{\{B_{10}\},\{B_{11}\},\{B_{12}, B_{14}\},\{B_{13}\},\{B_{15}\},\{B_{16}\},\{B_{17}\},\{B_{18}\},\{B_{19}\}\}$	0.727	11.70
		$\{\{B_{10}\},\{B_{11}\},\{B_{12}, B_{14}\},\{B_{13}\},\{B_{15}\},\{B_{16}, B_{19}\},\{B_{17}\},\{B_{18}\}\}$	0.699	27.69
		$\{\{B_{10}\},\{B_{11}\},\{B_{12}, B_{13}, B_{14}\},\{B_{15}\},\{B_{16}, B_{19}\},\{B_{17}\},\{B_{18}\}\}$	0.684	36.97
		$\{\{B_{10}\},\{B_{11}\},\{B_{12}, B_{13}, B_{14}, B_{15}, B_{16}, B_{19}, B_{17}, B_{18}\}\}$	0.582	55.98
		$\{\{B_{10}, B_{11}\},\{B_{12}, B_{13}, B_{14}, B_{15}, B_{16}, B_{17}, B_{18}, B_{19}\}\}$	0.536	81.03

Table 6.3 Best partitions of $\{T_{10}, T_{11}, \ldots, T_{19}\}$

α	Best partition (π)	δ	*Goodness* (π)
0.07	$\{\{T_{10},T_{13},T_{14},T_{16},T_{17}\},\{T_{11}\},\{T_{12},T_{15}\},\{T_{18},T_{19}\}\}$	0.725	85.59
0.06	$\{\{T_{10},T_{11},T_{15},T_{16},T_{17},T_{18}\},\{T_{12}\},\{T_{13},T_{14},T_{19}\}\}$	0.733	81.08
0.05	$\{\{T_{10}\},\{T_{11}\},\{T_{12}\},\{T_{13}\},\{T_{14},T_{16}\},\{T_{15}\},\{T_{17},T_{19}\},\{T_{18}\}\}$	0.890	13.35
0.04	$\{\{T_{10}\},\{T_{11},T_{13}\},\{T_{12}\},\{T_{14}\},\{T_{15}\},\{T_{16}\},\{T_{17}\},\{T_{18}\},\{T_{19}\}\}$	0.950	-2.07
0.03	$\{\{T_{10}\},\{T_{11}\},\{T_{12}\},\{T_{13}\},\{T_{14},T_{18}\},\{T_{15}\},\{T_{16}\},\{T_{17}\},\{T_{19}\}\}$	0.881	0.01

Table 6.4 Best partitions of $\{B_{10}, B_{11}, \ldots, B_{19}\}$

α	Best partition (π)	δ	*Goodness* (π)
0.020	$\{\{B_{10}\},\{B_{11},B_{12},B_{13},B_{14},B_{15},B_{16},B_{17},B_{18},B_{19}\}\}$	0.668	51.90
0.017	$\{\{B_{10}\},\{B_{11},B_{12},B_{13},B_{14},B_{15},B_{16},B_{17},B_{18},B_{19}\}\}$	0.665	66.10
0.014	$\{\{B_{10}\},\{B_{11},B_{12},B_{13},B_{14},B_{15},B_{16},B_{17},B_{18},B_{19}\}\}$	0.581	72.15
0.010	$\{\{B_{10},B_{11}\},\{B_{12},B_{13},B_{14},B_{15},B_{16},B_{17},B_{18},B_{19}\}\}$	0.560	63.67
0.009	$\{\{B_{10},B_{11}\},\{B_{12},B_{13},B_{14}\},\{B_{15}\},\{B_{16},B_{19}\},\{B_{17}\},\{B_{18}\}\}$	0.536	81.03

6.5 Conclusions

Clustering a set of databases is an important activity. It reduces cost of searching relevant information required for many problems. We provided an efficient solution to this problem in three ways. Firstly, we proposed more suitable measures of similarity between two databases. Secondly, we showed that there is a need to figure out

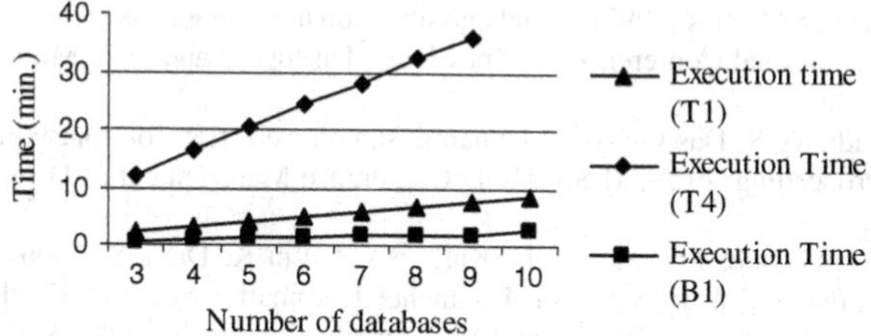

Fig. 6.3 Execution time vs. the number of databases

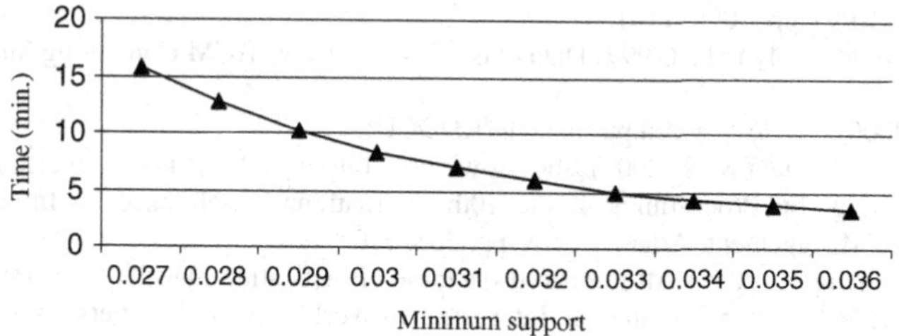

Fig. 6.4. Execution time vs. α for experiment with $\{T_{10}, T_{11}, \ldots, T_{19}\}$

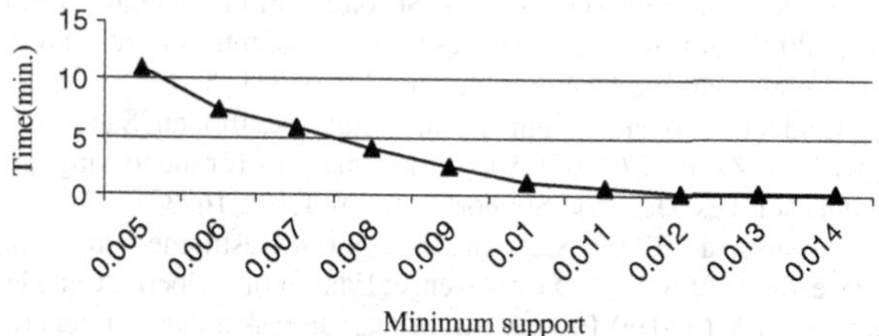

Fig. 6.5 Execution time vs. α for experiment with $\{B_{10}, B_{11}, \ldots, B_{19}\}$

the existence of the best clustering only at a few similarity levels. Thus, the proposed clustering algorithm executes faster. Lastly, we introduce IS coding for storing frequent itemsets in the main memory. It allows more frequent itemsets to participate in the clustering process. The IS coding enhances the accuracy of the clustering process. Thus, the proposed clustering technique is efficient in finding clusters in a set of databases.

References

Adhikari A, Rao PR (2008) Efficient clustering of databases induced by local patterns. Decision Support Systems 44(4):925–943

Agrawal R, Imielinski T, Swami A (1993) Mining association rules between sets of items in large databases. In: Proceedings of ACM SIGMOD Conference, Washington, DC, pp. 207–216

Ali K, Manganaris S, Srikant R (1997) Partial classification using association rules. In: Proceedings of the 3rd International Conference on Knowledge Discovery and Data Mining, Menlo Park, CA, pp. 115–118

Babcock B, Chaudhury S, Das G (2003) Dynamic sample selection for approximate query processing. In: Proceedings of ACM SIGMOD Conference Management of Data, New York, pp. 539–550

Bandyopadhyay S, Giannella C, Maulik U, Kargupta H, Liu K, Datta S (2006) Clustering distributed data streams in peer-to-peer environments. Information Sciences 176(14): 1952–1985

Barte RG (1976) The Elements of Real Analysis. Second edition, John Wiley & Sons, New York

FIMI (2004) http://fimi.cs.helsinki.fi/src/

Frequent Itemset Mining Dataset Repository (2004) http://fimi.cs.helsinki.fi/data

Huffman DA (1952) A method for the construction of minimum redundancy codes. In: Proceedings of the IRE 40(9), pp. 1098–1101

Jain AK, Murty MN, Flynn PJ (1999) Data clustering: A review. ACM Computing Surveys 31(3): 264–323

KDD CUP (2000) http://www.ecn.purdue.edu/KDDCUP

Lee C-H, Lin C-R, Chen M-S (2001) Sliding-window filtering: An efficient algorithm for incremental mining. In: Proceedings of the 10th International Conference on Information and Knowledge Management, Atlanta, USA, pp. 263–270

Li H, Hu X, Zhang Y (2009) An improved database classification algorithm for multi-database mining. In: Proceedings of the 3d International Workshop on Frontiers in Algorithmics, Springer, Berlin/Heidelberg, pp. 346–357

Ling CX, Yang Q (2006) Discovering classification from data of multiple sources. Data Mining Knowledge Discovery 12(2–3): 181–201

Liu CL (1985) Elements of Discrete Mathematics. Second edition, McGraw-Hill, New York

Liu H, Lu H, Yao J (2001) Toward multi-database mining: Identifying relevant databases. IEEE Transactions on Knowledge and Data Engineering 13(4): 541–553

Sayood K (2000) Introduction to data compression. Morgan Kaufmann, San Francisco

Su K, Huang H, Wu X, S. Zhang S (2006) A logical framework for identifying quality knowledge from different data sources. Decision Support Systems 42(3): 1673–1683

Tan P-N, Kumar V, Srivastava J (2002) Selecting the right interestingness measure for association patterns. In: Proceedings of SIGKDD Conference, Edmonton, Alberta, Canada, pp. 32–41

Wu X, Wu Y, Wang Y, Li Y (2005a) Privacy-aware market basket data set generation: A feasible approach for inverse frequent set mining. In: Proceedings of SIAM International Conference on Data Mining, pp. 103–114

Wu X, Zhang C, Zhang S (2005b) Database classification for multi-database mining. Information Systems 30(1): 71–88

Yang W, Huang S (2008) Data privacy protection in multi-party clustering. Data and Knowledge Engineering 67(1): 185–199

Yin X, Han J (2005) Efficient classification from multiple heterogeneous databases. In: Proceedings of 9-th European Conf. on Principles and Practice of Knowledge Discovery in Databases, pp. 404–416

Yin X, Yang J, Yu PS, Han J (2006) Efficient classification across multiple database relations: A crossmine approach. IEEE Transactions on Knowledge and Data Engineering 18(6): 770–783

Zhang S (2002) Knowledge discovery in multi-databases by analyzing local instances, Ph D thesis, Deakin University

Zhang T, Ramakrishnan R, Livny M (1997) BIRCH: A new data clustering algorithm and its applications. Data Mining and Knowledge Discovery 1(2): 141–182

Zhang S, Wu X, Zhang C (2003) Multi-database mining. IEEE Computational Intelligence Bulletin 2(1): 5–13

Chapter 7
A Framework for Developing Effective Multi-database Mining Applications

Multi-database mining has been already recognized as an important and strategically essential area of research in data mining. In this chapter, we discuss how one can systematically prepare data warehouses located at different branches for ensuring data mining activities. An appropriate multi-database mining technique is essential to develop efficient applications. Also, the efficiency of a multi-database mining application could be improved by processing more patterns in the individual application. A faster algorithm could also contribute to the enhanced quality of the data mining framework. The efficiency of a multi-database mining application can be enhanced by choosing an appropriate multi-database mining model, a suitable pattern synthesizing technique, a better pattern representation technique, and an efficient algorithm for solving the problem.

7.1 Introduction

More than 15 years have passed since Agrawal et al. (1993) introduced support-confidence framework for mining association rules in a database. Since then, there has been an orchestrated effort focused on a variety of ways of making the data mining in large databases as efficient as possible. With this regard, many interesting data mining algorithms (Agrawal and Srikant 1994; Coenen et al. 2004; Han et al. 2000; Toivonen 1996; Wu et al. 2004) have been proposed. But, the requirements and expectations of the users have not been fully satisfied. New and challenging applications arise over time. Multi-database mining applications are among those ongoing challenges.

Most of the existing algorithms have attempted to address ways of mining large databases. In this context, many parallel data mining algorithms (Agrawal and Shafer 1999; Chattratichat et al. 1997; Cheung et al. 1996) have been reported. These algorithms can be used to mine multiple databases by amalgamating them. It requires an organization to acquire parallel computing system. Such solution might not be suitable in many situations as these hardware requirements may easily result in quite significant and somewhat questionable investments.

In the context of mining multiple large databases we have discussed three approaches to mining multiple large databases (Chapter 1). In Section 7.2, we

A. Adhikari et al., *Developing Multi-database Mining Applications*, Advanced Information and Knowledge Processing, DOI 10.1007/978-1-84996-044-1_7,

discuss the shortcomings of these approaches. There are two categories of multi-database mining techniques. Some of them are specialized techniques, while remaining techniques are quite general in their nature. In Chapter 3, we presented the existing multi-database mining techniques. The choice of an appropriate multi-database mining technique becomes an important issue. When developing an efficient multi-database mining application there are several important components to be considered. There are many strategies using which one could develop a multi-database mining application. One should stress, though, that not all solutions could be equally efficient or suitable for the given application. The goal of this chapter is to offer a comprehensive framework to support the systematic development of multi-database mining applications.

A multi-database mining application can be developed through a sequence of several stages (phases) and each of these stages can be designed within its own framework. Thus an effective application can be developed by applying each stage in a systematic manner. In Section 7.3, we move on to a detailed discussion on different techniques aimed at the improvement of the process of multi-database mining applications. First we analyze why the existing approaches are not sufficient to develop an effective multi-database mining application.

7.2 Shortcomings of the Existing Approaches to Multi-database Mining

Let us briefly note that, as discussed in Chapter 1, there are three important approaches to multi-database mining such as local pattern analysis, sampling, and re-mining. To apply a multi-database mining technique, it is required to prepare the local databases. In the proposed framework, we wish to discuss this issue in details. Moreover, these techniques do not apply any optimization technique in the process of developing a multi-database mining application. We see later how one could apply such techniques to the development of an effective application. Again, these techniques do not talk about systematizing the development process of an application. We wish to stress on this issue also. One of the main hurdles we are faced with when dealing with multi-database mining applications that deal with mining multiple databases with high degree of accuracy. Moreover, synthesis of non-local patterns is a crucial stage for the first two approaches, while it remains a simple task for the third approach of mining multiple large databases. Unfortunately, the of *re-mining* approach is not advocated since it requires mining each of large databases twice.

7.3 Improving Multi-database Mining Applications

The main problem of multi-database mining is that it involves mining multiple large databases. Moreover, it is very likely that these databases might have been created without any coordination. We believe there is a need to systematize and improve the development stages of a multi-database mining application. We discuss various strategies for improving multi-database mining applications. Some improvements are general in nature, while others are more domain–specific. There are various techniques by which one could enhance the efficiency of multi-database mining applications. The efficiency of a multi-database application could be enhanced by choosing an appropriate multi-database mining model, a suitable pattern synthesizing technique, a better pattern representation technique and a more efficient algorithm to solve the problem. In addition, there are other important issues as discussed in the following sub-sections. In this book, we have illustrated each of these issues either in the context of a specific problem, or in a general setting. We do not stress much on efficient implementations of different algorithms, since this topic has been studied very intensively and is well-documented in the literature.

7.3.1 Preparation of Data Warehouses

As before, we consider an organization that has multiple databases at its different branches. It could well be that all the data sources are not of the same format. Many times data need to be converted from one type to another. One needs to process them before any mining task takes place. Relevant data are required to be retained for the purpose of mining. Also, the definitions of data are required to be the same at every data source. The preparation of data warehouse completed at every branch of the organization could be a significant task (Pyle 1999; Zhang et al. 2003). We have presented an extended model (in Chapter 2) for synthesizing global patterns from local patterns in different databases. We have discussed how this model could be used for mining heavy association rules in multiple databases. Also, it has been shown how the task of data preparation could be broken into sub-tasks so that the overall data preparation task becomes easier and can be realized in a systematic fashion. Although the above model introduces many layers and interfaces for synthesizing global patterns, many of these layers and interfaces might not be required in a real-life application. Due to the heterogeneous nature of different data sources, data integration is often one of the most challenging tasks in managing modern information systems. Jiang et al. (2007) have proposed a framework for integrating multiple data sources when a single "best" value has to be chosen and stored for every attribute of an entity.

7.3.2 Choosing Appropriate Technique of Multi-database Mining

Zhang et al. (2003) designed local pattern analysis for mining multiple large databases. It returns approximate global patterns in multiple large databases. In many multi-database mining analyses, local pattern analysis alone might not be sufficient. Thus, one might need different techniques in different situations. A certain technique of mining multiple databases could not be appropriate in all situations. Its choice has to be implied by the problem at hand. We have presented a multi-database mining technique, MDMT: PFM+SPS, for mining multiple large databases (see

Chapter 3). It improves multi-database mining when being compared with an existing technique that scans each database only once. Experimental results in Chapter 3 have shown the effectiveness of this technique. It has to be noted, though, that it does not mean that such algorithm is the best in all situations. For example, we have presented a technique for mining multiple large databases to study problems involving a set of specific items in multiple databases (Chapter 4). It happened to perform better than the MDMT: PFM+SPS. It extracts true patterns related to a set of specific items coming from multiple databases. The multi-database mining presented in Chapter 4 is an important as well as highly promising issue, since many data analyses of a multi-branch company are based on select items. The choice of a multi-database mining technique is an important design issue.

7.3.3 Synthesis of Patterns

As discussed in Chapter 3, a multi-database mining using local pattern analysis is two-step process. At the first step we apply a model for mining each local database using a SDMT. We synthesize non-local patterns using local patterns in different databases at the second stage. In many applications (Adhikari and Rao 2008; Wu and Zhang 2003), the synthesis of patterns is an important component. It is always better to avoid the stage of synthesizing patterns. For example, while mining global patterns of select items in multiple databases, we have adopted a different technique (Fig. 4.1). In this case the chosen multi-database mining technique does not require the synthesizing step and returns true global patterns of select items. In fact, in this technique there is no need to synthesize patterns. In many applications, it might not be possible to avoid the synthesizing step. In such situations, one needs to apply a multi-database mining technique that returns high quality of patterns. In Chapter 3, we have presented one such technique, namely MDMT: PFM+SPS.

7.3.4 Selection of Databases

For answering a query, one needs to select appropriate databases. Their selection is based on the inherent knowledge residing in the database. One needs to mine each of the local databases. Then we process the local patterns in different databases for the purpose of selecting relevant databases. Local patterns help selecting relevant databases. Based on local patterns, one can cluster the local databases. For answering the given query, one mines all the databases positioned in a relevant cluster. In many cases, the clustering of databases is based on a measure of similarity between these databases. Thus, the measure of similarity between two databases is an important design component whose development is based on local patterns present in the databases.

Wu et al. (2005) have proposed a similarity measure sim_1 to identify similar databases based on item similarity. The authors have designed an algorithm based on this measure to cluster databases for the purpose of selecting relevant databases.

Such clustering is useful when the similarity is based on items present in different databases. This measure might not be useful for many multi-database mining applications where clustering of databases might be based on some other criteria. For example, if we are interested in the relevant databases based on transaction similarity then the above measures might not be appropriate. We have presented a technique for clustering databases based on transaction similarity (Chapter 6). We have introduced a similarity measure $simi_1$ to cluster different databases and designed a clustering algorithm based on $simi_1$ for the purpose of selecting relevant databases.

An approximate form of knowledge resulting from large databases would be adequate for many decision support applications. In this sense, the selection of databases might be important in many decision support applications by reducing the cost of searching for necessary information.

7.3.5 Representing Efficiently Patterns Space

Usually an application dealing with multiple databases often handles a large number of patterns. Multi-database mining using local pattern analysis is an approximate method of mining multiple large databases. One needs to improve the quality of knowledge synthesized from multi-database mining. The quality of synthesized global patterns or a decision based on local patterns could be enhanced by incorporating more local patterns in the knowledge synthesizing/processing activities. One could incorporate more local patterns by using a suitable coding technique. Frequent itemset and association rule are two important and interesting types of pattern in a database. In the context of storing patterns space efficiently, we have presented two coding techniques:

7.3.5.1 Representing Association Rules

Association rule mining (ARM) has received a lot of attention in the KDD community. Accordingly, many algorithms on ARM have been reported in the recent time. We have observed that the number of association rules generated from a moderate-size database could be quite large. Therefore an application that mines multiple large databases and applies local pattern analysis often handles a large number of association rules. To develop an effective application, we have presented the ACP coding to represent association rules in multiple databases space efficiently (Chapter 5). Such applications improve the quality of synthesized global association rules. We have included experimental results to show the effectiveness of ACP coding for representing association rules in multiple databases.

7.3.5.2 Representing Frequent Itemsets

In the process of extracting association rules in a database, one needs to extract frequent itemsets from the database. In many applications, frequent itemsets are used to find the solutions. As noted in the previous section, a multi-database mining application often handles a large number of frequent itemsets. To improve the quality of the application one needs to incorporate a large number of frequent itemsets. In view of this objective, we have presented the IS coding to represent frequent itemsets in local databases space efficiently (Chapter 6). The theoretical analysis quantifies the effectiveness of this coding.

7.3.6 Designing an Appropriate Measure of Similarity

Many algorithms are based on a measure used for decision making. For example, most of the clustering algorithms are based on a measure of association. Such clustering algorithms become more accurate if the similarity measure used in an algorithm becomes more appropriate towards measuring the similarity between two objects under consideration. For example, if we are interested in mining association patterns approximately in multiple large databases, then the information regarding the association among items would be available in itemsets rather than in data items in different data sources (Chapter 6). In this case, a measure based on itemsets in different data sources seem to be more appropriate in finding similarity between two databases. The efficiency of a clustering algorithm is dependent on the suitability of the similarity measure used in the algorithm.

7.3.7 Designing Better Algorithm for Problem Solving

Using suitable data structures and the algorithm one supports the realization of the efficient multi-database mining applications (Aho et al. 1987; Aho et al. 1974). In the context of extracting high-frequent association rules in multiple databases, we have designed an algorithm that runs faster than the existing algorithms (Chapter 2). Moreover, our algorithm is simple and straightforward. In the context of clustering the databases, we have designed an improved algorithm based on different parameters (Chapter 6). In this algorithm, we have enhanced efficiency of the clustering process using the following strategies: We use more appropriate measure of similarity between two databases. Also we determine the existence of the best clustering only at few similarity levels. Thus, the clustering algorithm executes faster. As the IS coding for storing frequent itemsets space is efficient, more frequent itemsets can participate in the clustering process. Thus, it makes the clustering process more accurate.

7.4 Conclusions

Multi-database mining applications might come with different complexities across different domains. It is difficult to establish a generalized framework for the development of efficient multi-database mining applications. Nevertheless, we can identify some important stages of the development process that are crucial to the overall performance of the data mining environment. The sound design practices supporting the phases identified in this chapter are essential to enhance the quality of many multi-database data mining applications.

References

Adhikari A, Rao PR (2008) Synthesizing heavy association rules from different real data sources. Pattern Recognition Letters 29(1): 59–71

Agrawal R, Imielinski T, Swami A (1993) Mining association rules between sets of items in large databases. In: Proceedings of ACM SIGMOD Conference, Washington, DC, pp. 207–216

Agrawal R, Shafer J (1999) Parallel mining of association rules. IEEE Transactions on Knowledge and Data Engineering 8(6): 962–969

Agrawal R, Srikant R (1994) Fast algorithms for mining association rules. In: Proceedings of International Conference on Very Large Data Bases, pp. 487–499

Aho AV, Hopcroft JE, Ullman JD (1974) The Design and Analysis of Computer Algorithms. Addison-Wesley, Reading, MA

Aho AV, Hopcroft JE, Ullman JD (1987) Data Structures and Algorithms. Addison-Wesley, Reading, MA

Coenen F, Leng P, Ahmed S (2004) Data structure for association rule mining: T-trees and P-trees. IEEE Transactions on Knowledge and Data Engineering 16(6): 774–778

Chattratichat J, Darlington J, Ghanem M, Guo Y, Hüning H, Köhler M, Sutiwaraphun J, To HW, Yang D (1997) Large scale data mining: Challenges, and responses. In: Proceedings of the Third International Conference on Knowledge Discovery and Data Mining, pp. 143–146.

Cheung D, Ng V, Fu A, Fu Y (1996) Efficient mining of association rules in distributed databases. IEEE Transactions on Knowledge and Data Engineering 8(6), 911–922.

Han J, Pei J, Yiwen Y (2000) Mining frequent patterns without candidate generation. In: Proceedings of ACM SIGMOD Conference on Management of Data, Dallas, TX, pp. 1–12

Jiang Z, Sarkar S, De P, Dey B (2007) A framework for reconciling attribute values from multiple data sources. Management Science 53(12): 1946–1963

Pyle D (1999) Data Preparation for Data Mining. Morgan Kufmann, San Francisco

Toivonen H (1996) Sampling large databases for association rules. In: Proceedings of the 22nd International Conference on Very Large Data Bases, pp. 134–145

Wu X, Zhang S (2003) Synthesizing high-frequency rules from different data sources. IEEE Transactions on Knowledge and Data Engineering 14(2): 353–367

Wu X, Zhang C, Zhang S (2005) Database classification for multi-database mining. Information Systems 30(1): 71–88

Wu X, Zhang C, Zhang S (2004) Efficient mining of both positive and negative association rules. ACM Transactions on Information Systems 22(3): 381–405.

Zhang S, Wu X, Zhang C (2003) Multi-database mining. IEEE Computational Intelligence Bulletin 2(1): 5–13

Index

A. Adhikari et al., *Developing Multi-database Mining Applications*, Advanced Information and Knowledge Processing, DOI 10.1007/978-1-84996-044-1,